Propagation

Alan Toogood
Propagation

J M Dent & Sons Ltd London Melbourne Toronto

First published 1980
© Alan Toogood 1980

Printed in Great Britain by
Billing & Sons Ltd, Guildford, London, Oxford, Worcester for
J M Dent & Sons Ltd, 33 Welbeck St, London
This book is set in 11/13 VIP Baskerville by
D. P. Media Limited, Hitchin, Hertfordshire

British Library Cataloguing in Publication Data
Toogood, Alan R
    Propagation.
    1. Plant propagation
    I. Title
    635'.04'3   SB119

    ISBN 0-460-04377-3

Contents

List of figures   6
List of plates   8
Acknowledgements   10
Introduction   11
1 The ancient art of propagation   13
2 Why propagate?   18
3 The fascination of propagation   21
4 Raising hardy plants from seeds   25
5 Raising tender plants from seeds   53
6 Stem cuttings   62
7 Other types of cuttings   83
8 Grafting   103
9 Budding   119
10 Division of plants   125
11 Simple layering   144
12 Other methods of layering   150
13 Nursery management   165
14 A guide to the alphabetical tables   176
   Table: Propagation of hardy garden plants   180
   Table: Propagation of greenhouse and other tender plants   268
   Select bibliography   305
   General index   306
   Index of plant names   307

List of figures

1 Storing maple (*Acer*) seeds in peat
2 Preparing seeds of holly (*Ilex*) for stratification
3 Sowing seeds of trees and shrubs in raised beds outdoors
4 Sowing seeds of alpines
5 Preparing a seed tray for sowing
6 Sowing seeds in a tray
7 Pricking out seedlings
8 Preparation of softwood shrub cuttings
9 Softwood cuttings being dipped in a hormone rooting powder
10 Inserting softwood shrub cuttings
11 Preparing semi-ripe cuttings of *Pelargonium zonale*
12 Mallet cuttings of *Berberis thunbergii* cultivar
13 Cuttings of a *Chamaecyparis lawsoniana* cultivar
14 Rooting semi-ripe shrub cuttings under a low polythene tunnel
15 Hardwood cuttings of black currant and red currant
16 Inserting hardwood cuttings in the open ground
17 Preparation of root cuttings
18 Inserting root cuttings
19 Leaf cuttings of *Saintpaulia*
20 Leaf cutting of *Begonia rex*
21 Leaf cuttings of *Sansevieria trifasciata*
22 Leaf-bud cuttings of rubber plant (*Ficus elastica*)
23 Eye cuttings of grape vine (*Vitis*)
24 Inserting eye cuttings
25 Pipings of pinks (*Dianthus*)
26 Whip-and-tongue grafting
27 Saddle grafting a *Rhododendron*
28 The spliced side graft
29 The veneer graft
30 The spliced side veneer graft

31 Rose budding
32 Dividing a herbaceous plant
33 Divisions of German or bearded *Iris*
34 *Sempervivum* offsets
35 Dividing a waterlily (*Nymphaea*)
36 Removing bulblets
37 Bulbils on stem of lily (*Lilium*)
38 Propagating a lily (*Lilium*) from scales
39 *Gladiolus* cormlets
40 Dividing a *Dahlia*
41 Simple layering a *Rhododendron*
42 Air-layering a *Rhododendron*
43 Air-layering a rubber plant (*Ficus elastica*)
44 Serpentine layering a *Clematis*
45 Tip layering a blackberry
46 Layering strawberry runners
47 Layering a border carnation
48 Layering a *Chlorophytum*
49 Potting
50 Pruning a young shrubby dogwood (*Cornus*)

List of plates

between pages 96 and 97

1 Cones of *Cedrus, Abies, Sequoiadendron* and *Pinus*
2 Seed heads of *Clematis, Corylus* and *Laburnum*
3 Seed heads of *Cercis, Piptanthus* and *Hydrangea*
4 *Cotoneaster* berries, *Ceanothus* seed heads and *Pyracantha* berries
5 Rose heps and *Leycesteria* seed heads
6 Seed heads of *Papaver, Lathyrus* and *Crocosmia*
7 Seed heads of *Kniphofia* and *Acanthus*
8 Seed heads of *Eryngium, Achillea* and *Lupinus*
9 Seed heads of *Phlomis, Limonium* and *Malva*
10 Grass seed heads
11 Seed heads of *Galtonia, Allium* and *Agapanthus*
12 Seed heads of *Incarvillea, Callistemon* and *Iris*
13 Collecting seeds of *Sorbus*
14 Collecting seed pods of *Lilium*
15 Use of a seed sieve
16 Mixture of seeds and chaff
17 Separating seeds and chaff

between pages 144 and 145

18 Preparing a seed bed
19 Raking a seed bed
20 Drills for seed sowing
21 Sowing seeds of hardy annuals
22 Covering seeds with soil
23 Seedlings in need of thinning
24 Seedlings thinned out
25 Fern frond with spore cases
26 Prothallus stage of fern propagation

27 Young ferns ready for potting
28 A newly potted young fern
29 An electrically heated propagating case
30 Mist-propagation unit
31 *Juniperus* cuttings rooting in cold frame
32 Cuttings rooting on windowsill indoors
33 Cuttings rooting in water
34 A small home nursery

Acknowledgements

For the illustrations: Mr R. Sweetinburgh

For the photography: Mr Dick Robinson of the Harry Smith Horticultural Photographic Collection

For valued advice and guidance: Mr Graham Stuart Thomas, Mr Arthur Turner.

Introduction

I had been considering writing a book on plant propagation for several years when the opportunity came to write this one. The publishers required a really comprehensive book on the subject for amateur gardeners – exactly what I had in mind – and this is the result. The aim of the book is to cover all the aspects of plant propagation that the amateur gardener is likely to require, and to list the propagation methods for a wide range of hardy and tender plants; indeed the tables contain over 1500 different kinds of plants. This is probably the most comprehensive propagation list in any modern amateur book on the subject. Horticultural students studying at the many colleges throughout the country should also find the tables useful, as should propagators employed by nurseries or in private gardens.

I have endeavoured to include all the up-to-date methods of propagation which might be used by amateur gardeners – for instance, rooting cuttings under low polythene tunnels. I have also included many useful hints obtained from the commercial world of plant propagation and which the amateur could well adopt, such as the use of rubber budding patches for rose budding; mallet cuttings for *Berberis*; and 'wounding' cuttings to increase the chances of rooting. Many of these tips, which originate from the nursery industry, are not usually mentioned in books for the amateur yet there is no reason at all why they should not be in more general use, and so increase the propagator's rate of success.

Methods of propagation need illustrating well and precisely, so that the gardener is left in no doubt about how to carry out the various operations. I have chosen to use mainly line drawings in this book – they are inclined to show propagation methods far more clearly than photographs – and I am most grateful to my talented colleague Roger Sweetinburgh for agreeing to draw the illustrations for this book.

Introduction

I am always aware that gardeners are forever seeking further information and that there are bound to be items in this book about which some readers would like to know more. There are various ways of obtaining information: the local library is one of the obvious ways, or perhaps writing to gardening magazines. Most have a readers' advisory service. The Royal Horticultural Society also has an advisory service based at its Garden at Wisley, near Ripley, Surrey. I have included a short book list which should prove useful, particularly those books which deal with the identification of plants. In a book of this length and type one cannot possibly include identification characteristics of all the plants, although in the tables I have indicated the type of plant in every case, such as shrub, tree, cactus, or whatever.

Gardeners who may not, perhaps, have attempted propagation before may well ask the question: why should I want to increase my plants? May I direct these readers to Chapters 2 and 3 which contain some extremely good reasons for taking up the gentle art of plant propagation.

Alan Toogood, Guildford, Surrey, 1979

1 The ancient art of propagation

The propagation of plants is as old as civilization itself. The sowing of seeds to produce food crops was undoubtedly the earliest form of propagation and indeed marked the start of the agricultural revolution many thousands of years ago, when man progressed from hunter and gatherer to grower and farmer. Man no doubt learned this natural method of increase by observing the dispersal of seeds by the plants around him and the subsequent emergence of seedlings.

Many of the ancient civilizations grew a good range of food crops, including cereals, from seeds and a very early written account is to be found in the book *Enquiry into Plants* by Theophrastus. The author was born in 370 BC at Eresos in Lesbos, Greece, and was a pupil of Plato and Aristotle. In his book we learn that the olive, date palm and cypress were grown from seeds, together with various 'herbaceous plants and pot herbs' like cabbage, radish, turnip, beet, lettuce, coriander, dill, cress, leeks, celery, orach, cucumber, basil, purslane, savory and marjoram.

At that period most seeds were sown in the open, but the Romans knew the advantages of 'glass' (thin sheets of mica) and heat for protecting and forcing fruit and vegetables, and also for germinating seeds of certain plants such as cucumbers. The Greeks may also have sown seeds in protected conditions, using a form of bell glass over seeds in pots or dishes, which were then placed out in the sun. Today we often use electrically heated propagation cases for the same purpose.

Other methods of plant propagation in use today are far from new: techniques like rooting cuttings, layering and grafting are ancient methods of increasing plants. The first mentioned is considered to be the most ancient method of vegetative propagation.

Again, Theophrastus wrote about taking cuttings of plants. The method of planting pieces of shoot with some roots attached was practised wherever possible. These shoots were rooted suckers which were pulled off the plants. Nowadays when we pull off a rooted shoot from a plant we call it an 'Irishman's cutting'. Some plants were propagated from unrooted shoots, such as the fig, olive, pear, apple, pomegranate, bay, almond, myrtle, rose, cypress, rue, marjoram and basil. I rather like the way in which Theophrastus described the propagation of plants: for example, 'from a piece torn off'; or in the case of the vine, 'grown from its branches'. The olive was also grown 'from a root' – presumably a sucker. Some of the vegetables and herbs mentioned above, as well as being raised from seeds, were also propagated 'by a piece torn off, a shoot, or a piece of root'. Celery, onions and leeks were also propagated by offsets.

It is interesting that Theophrastus observed that the seed-raised offspring of cultivated fruits had different characteristics to the parent plants, and therefore they were propagated vegetatively. He noted that there was a degeneration of quality when cultivated fruits were seed-raised.

At the time of Theophrastus there was no sophisticated propagation equipment; cuttings, suckers and so on were generally planted directly in the field, but 'glass' and heat from the sun may have been used to encourage quicker growth.

An interesting aspect of taking cuttings comes from the Romans: apparently the bases of cuttings were dipped in ox manure before insertion. This was no doubt to stimulate rooting and encourage a strong root system; I am glad that today we use hormone rooting powders instead, to ensure good rooting in our cuttings.

It seems that the Romans may have been the first to propagate plants from root cuttings for we understand that a plant something like a thistle was increased by this method. However there is not much evidence of propagation from root cuttings until the end of the seventeenth century, at which time citrus trees particularly were increased from pieces of root.

The Victorians had a novel way of rooting cuttings. They placed a small pot inside a larger one and filled the space in between with compost in which the cuttings were inserted. Water was poured into the inner pot to moisten the compost. Over this arrangement was placed a bell glass to maintain a humid atmosphere around the cuttings. Bell glasses were, in fact, used well into this century for propagation and the forcing of young plants but now they are scarce and considered collectors' items. Bell glasses were the first type of cloche, and they were indeed bell shaped, made of clear glass and with a knob at the top for picking them up. There are modern plastic equivalents of bell glasses which are useful for covering pots of cuttings, either in a greenhouse or on a windowsill indoors.

Bottom-heat to speed the rooting of cuttings was used by Victorian gardeners and nurserymen and they found it particularly useful for rooting cuttings of the great range of tropical plants that were being brought to Britain by plant collectors at the time. Propagation cases with glass covers or lids were often used and paraffin lamps below provided the heat. Since then we have advanced considerably and there are now electrically heated propagating cases in a wide range of sizes. Also many nurserymen, and private gardeners, use electric soil-warming cables bedded in sand or shingle on a greenhouse bench to provide bottom-heat for propagation purposes. The most sophisticated equipment is the mist-propagation unit which was introduced from America in the 1950s. Here the cuttings are automatically sprayed intermittently with a fine mist-like spray of water to prevent them from wilting, and they are warmed from below by electric soil-warming cables.

We tend to think of grafting as being an artificial form of propagation but in fact it can occur in nature. If the branches of two trees, or the same tree, constantly rub together so that each is wounded, for example, then it is possible that the branches will unite or grow together to form a permanent bond. Furthermore, if young stems are in very close contact with each other they may unite. It is fairly common to see natural grafts in beech (*Fagus sylvatica*), elm (*Ulmus procera*), ash (*Fraxinus excelsior*), common maple (*Acer campestre*), poplar (*Populus*) and ivy (*Hedera helix*). Roots can only unite to form natural grafts: either the roots of a single tree or the roots of neighbouring trees.

It is likely that man, in the distant past, observed natural grafting in the plants around him and so came upon the idea of joining together parts of two separate plants to form a new plant.

Modern grafting involves joining a selected variety (using detached scions) on to a rootstock – generally the common counterpart of the variety. However, it seems probable that the very earliest type of grafting was approach grafting, where the branch of one tree was securely fixed to the branch of another tree, after cuts had been made in the bark of each branch.

It is known that the ancient Greeks carried out grafting with detached scions; and also the Romans, especially with roses which they found developed slowly from seeds. It was the Romans who began to realize that a rootstock can influence the vigour and size of a tree. They used a number of methods of grafting including the cleft graft, where a scion or shoot is inserted into a wedge made between the bark and wood of the rootstock. It is possible that the Romans also grafted a single rootstock with a number of different varieties of apple (or other fruit) to produce what we call today the family tree, very popular with small-garden owners or those with limited space.

The ancient civilizations even protected their grafts from moisture and diseases by sealing them. We know that they used a mixture of clay and chalk together with sand, cattle manure and straw. The modern method of sealing grafts is to use plastic grafting tape, grafting wax or bituminous tree paint.

The Romans also propagated plants by budding, placing a bud into a cut on the rootstock. Today this is a common method of propagating roses and many kinds of trees, both fruiting and ornamental.

The ancient Chinese knew of the desirability of using rootstocks for budding. Indeed they used the vigorous wild tree peony or moutan as a rootstock for budding varieties of this plant. They produced many varieties of tree peony by hybridization, particularly during the eighth and ninth centuries, and increased them vegetatively by budding.

The propagation of plants by grafting increased tremendously in the sixteenth and seventeenth centuries and there were many more kinds of graft than are in use today. In fact, nurserymen have in recent times purposely reduced the number of grafts to simplify somewhat this method of plant increase. In the eighteenth century family trees, especially of citrus fruits, were produced by a novel method, which involved grafting together a number of rooted cuttings of different varieties. When these had united they were generally potted into a container of some kind.

Also in the eighteenth century there was much natural grafting of trees of the same species. This was carried out when the trees were young and pliable and often they were twisted and tied together into various shapes – for instance, they would be formed into the shape of pillars or balustrades. What a pity there seems little time, or inclination, these days for such artistic work.

It is possible that natural layering of wild plants gave man the idea of encouraging branches to form roots while still attached to the parent plant. Indeed, in the countryside one can often observe the branches of trees, or the stems of brambles, for instance, rooting into the soil. We know that the Romans practised layering of vines, for it was written about by Columella, a Roman writer on agriculture in the first century BC. The modern method of layering involves wounding the part of the shoot that is to be buried in the soil, but the Romans did not believe in this as they thought it would weaken it.

Air-layering, which involves rooting a shoot above ground level while it is still attached to the parent plant, was, we understand, invented by the ancient Chinese and indeed many people know this technique as Chinese layering. In Roman times Cato (234–149 BC) described this method of layering and it was also probably used by the ancient Greeks to propagate trees. In early times the shoot was rooted in a pot or basket filled with soil. Of course this method can be used successfully today but the modern approach is to wrap part of the shoot with sphagnum moss, this being held in place with a sleeve of clear polythene.

So the methods of propagation that we use today are not new: they have simply been modified or improved. It is only in this century that completely new methods of plant propagation have been discovered, like tissue culture, using minute pieces of plant tissue to produce new plants, under completely sterile laboratory conditions. This method is being taken up by some nurserymen, especially orchid producers, and enables the propagator to obtain a great many new plants from a single parent plant. It is also used in the production of virus-free plant material, such as fruit trees, and a number of research stations are using this technique to 'clean up' a wide range of fruit. The extreme tips of young shoots are taken as these do not contain virus. Tissue culture (also known as meristem propagation) is not yet within the scope of the amateur gardener, however.

The production of seedlings under completely artificial conditions of light and heat is now being used by some nurserymen, especially the large producers of summer bedding plants. The seedlings are raised in cabinets called 'growing rooms' and they never see the light of day until they have reached a certain stage of development. The advantage of this method is that the environment is accurately controlled and never varies in respect of light and temperature; this ensures optimum germination of seed and very even or uniform growth in seedlings. There is also a great deal of research being carried out in the use of artificial light for the rooting of cuttings and the growing on of the resultant young plants.

2 Why propagate?

Most amateur gardeners do not propagate plants indiscriminately but have definite and very good reasons for wishing to increase their stocks. So let us now consider some of the major reasons for producing our own plants.

First let us look at costs. Plants are becoming ever more expensive to buy and therefore it is very much cheaper to raise your own wherever possible. In no way do I wish to criticize nurserymen about the high cost of plants, for prices are carefully calculated to ensure that nursery businesses, garden centres and so on run at a profit, but at the same time give the customers as fair a deal as possible. It costs a great deal of money nowadays to produce plants on a nursery, what with the cost of labour, oil for heating glasshouses and all the sundries that are required, such as pots, composts and fertilizers. Plants are still good value for money, especially long-lived subjects like trees, shrubs, roses and conifers. But even so, many people just cannot afford to buy plants in quantity. With a knowledge of propagation, however, it is quite possible to produce batches of young plants very cheaply.

For instance, with a heated greenhouse all your summer bedding plants can be raised from seeds. Many other plants are easily raised from seeds too, such as trees and shrubs, rock plants or alpines, hardy perennials or herbaceous plants, hardy biennials like wallflowers (*Cheiranthus*), hardy annuals and even bulbs. A wide range of pot plants can also be grown from seeds. There is no doubt that seeds are still the cheapest means of obtaining new plants – packets still only cost a matter of pence rather than pounds.

Many, if not all, of the existing plants in the garden and greenhouse can be increased in various ways if you possess a knowledge of propagation. All kinds of hardy and greenhouse plants can be propagated from cuttings of various types, for instance. It is not even necessary to have a heated greenhouse to root some types of cuttings; a cold frame or even a sheltered spot out of doors are quite adequate for the rooting of many kinds of plants.

Other ways of increasing existing plants are budding, grafting, dividing and layering. New roses can be produced by budding, and also some favourite trees. Trees can be produced by grafting, too. Many plants can be increased very easily by dividing or splitting them into smaller portions – hardy perennials or herbaceous plants, for example, and also alpines, greenhouse pot plants and even some shrubs and bulbous subjects. Finally there is layering, which is a very easy method of raising new trees, shrubs, fruits and even certain greenhouse subjects.

All of these methods of propagation are explained in detail in the following chapters, and the second part of the book consists of comprehensive tables which indicate the most suitable methods of propagation for over 1500 different kinds of plants.

Home propagation really comes into its own when a good quantity of expensive plants is wanted. Those which come to mind immediately, of course, are spring and summer bedding plants for they are required each year. They have to be planted fairly close together for the best effect, and this means a lot of plants are required even for small beds and borders.

Trees and shrubs are often expensive to buy, especially choice genera like *Rhododendron*, *Magnolia*, *Hamamelis* (witch hazel) and *Camellia*, and therefore most people buy just one plant. With a knowledge of propagation, however, it is comparatively easy to increase such plants so that eventually you are able to plant them in groups: this, of course, creates far more impact than single specimens. Other plants that you may wish to plant in bold groups include herbaceous plants and alpines; and again, if single specimens are bought stocks can be increased very easily and cheaply.

But enough about costs – once you start propagating you will very quickly realize how much money can in fact be saved. Another good reason for propagation is to build up stocks to give away as presents to gardening friends and relations. Plants make ideal gifts for anyone interested in gardening and it often means so much more to the recipients if they realize that you have raised the plants. If you are a member of a gardening club or society then surplus plants are always useful to take along to the plant stall which many clubs have at their meetings. These sales stalls are often very popular with members and help to raise money for the club funds.

Why propagate?

It is also useful to have some spare plants of the subjects which you grow in your garden or greenhouse so that you are able to exchange items with friends or neighbours. This in fact is a very good way of building up a collection of plants. Then in your travels you may be given cuttings or seeds of some particular plant that you admire in a garden. With a knowledge of plant raising there is a far better chance of raising new plants from such material. There is nothing more frustrating than to dib in some cuttings or sow seeds which you have been given only to find that they die or do not germinate.

Many garden plants, especially old-fashioned cultivars, have been saved from extinction over the years (particularly over the war years) by amateur propagators – plants like old-fashioned pinks (*Dianthus*), auriculas (*Primula*), violas and so forth. Nurserymen cannot possibly propagate all the plants which are in cultivation, especially the great wealth of cultivars and hybrids, and so it is often up to the amateur gardeners to keep stocks going. It could almost be said that it is the duty of the amateur gardener to propagate some of the plants he or she possesses in order to ensure that they remain in cultivation. Many highly developed plants that we grow today, especially hybrids with double flowers, will not reproduce naturally from seeds and therefore if they are not propagated by vegetative means they stand no chance of naturally perpetuating themselves.

So here is a great incentive for the amateur gardener – very choice or old-fashioned plants should be high on the list. Propagate plants and distribute to as many friends and other people as possible, to ensure that no subjects are lost to cultivation.

In your garden you may well find that you have some plants which have a fairly short productive life before they start to decline in vigour and fade away. For instance, carnations and pinks (*Dianthus*), strawberries (*Fragaria*), lupins (*Lupinus*), *Delphinium*, heaths and heathers (*Erica* and *Calluna*), *Aquilegia* and *Meconopsis* all need replacing with new plants on a regular basis, and if you are able to raise your own replacements then this saves you the expense of buying. Very often, of course, you may not bother to buy new plants so that eventually your plants die and you are without those particular subjects. So here is another very good motive for propagation.

I hope that I have given some worth-while reasons for home propagation. Another very good argument for raising your own plants is discussed in the next chapter – and this is the pure fascination of the subject. What better justification could there possibly be for propagation?

3 The fascination of propagation

Propagation is a fascinating subject and is well worth taking up for the sheer interest and enjoyment that it provides. It will certainly widen your interest in plants and makes a refreshing change from the straight-forward garden maintenance which all garden owners are obliged to undertake to ensure that plants thrive and the garden looks present-able.

You can really get to know plants when you propagate them. You will find that you look at plants more closely, in greater detail than ever before. You will observe their habits and characteristics and the way they develop from seedlings, cuttings and so on into full-sized plants. All too often, when you simply grow plants, you do not really study them in great detail. You may grow ferns in your garden or greenhouse, for instance. Did you realize that on the backs of the fronds there are many millions of minute 'spores' – the reproductive bodies, or the 'seeds' of the fern world? Ferns can be raised from these spores, as described in Chapter 4.

Did you know that some plants reproduce themselves vegetatively, by producing young plants on their leaves? Some ferns do this, as do *Bryophyllum* and *Tolmiea*, and the tiny plantlets will grow into perfect replicas of the parent plants.

Collecting seeds from your garden and greenhouse plants will open your eyes to a wide variety of seed pods and capsules that are produced by plants– they come in an amazing range of shapes and sizes, as can be seen in Plates 1–12. Many seed pods have the habit of exploding when the seeds are ripe, scattering the seeds far and wide. Collecting these seeds before they are shed calls for very close observation of the plant!

Most gardeners know that many plants can be propagated from ordinary stem cuttings, but I wonder how many people realize that numerous plants will grow from roots cut into small sections. These are known as root cuttings. Then many plants can also be propagated from leaves – either complete leaves or sections of leaves – which will produce plantlets that will grow and develop into specimens which resemble their parents in every respect.

Venturing into grafting gives the fascination of making a new plant by permanently joining together parts of two separate plants. You will learn about the 'cambium layers' – the tissue underneath the bark of plants which fuses together the two separate portions of plants.

When it comes to the propagation of bulbs, you will realize that the actual bulbs are composed of 'scales', which are modified swollen leaves. Some lilies (*Lilium*) are propagated by detaching and planting these scales. Other bulbs produce tiny bulbs on their flower stems and if these are detached and planted they will develop into full-sized flowering plants.

Another really fascinating aspect of propagation is the rooting of stems and shoots while they are still attached to the parent plant. This is known as layering. In many cases the shoots are rooted in the soil, but sometimes shoots can be rooted well above ground level and this is known as 'air-layering'.

There are many skills to be mastered in propagation. With practice you will become expert in the use of a knife – either a budding knife or a grafting type – and this can give great satisfaction. In time you will not be content to make ragged cuts with perhaps a blunt knife, which invariably results in failure of cuttings to root or grafts to 'take', but you will be endeavouring to make really neat clean cuts with a proper horticultural knife.

Pricking out or transplanting tiny seedlings is another skill that can be perfected with practice. The smaller the seedlings when they are transplanted the greater the chance they will establish and grow into strong healthy plants.

Then there is the art of potting young plants, rooted cuttings and so on to learn – there is more to this than at first meets the eye. You will discover how to plant out young plants correctly, to ensure that they become established and grow, and of course some plants will need training to ensure they develop into acceptable specimens. This will call for the skilful use of secateurs at the optimum stages of the plants' development. Careful training is most important when you produce new trees by grafting or budding: you will have to decide what form your tree is to take – for instance, a standard, half standard or perhaps a bush tree. Fruit trees can also be trained to many other shapes – such as cordons, fans and espaliers. Then many shrubs need to be pruned correctly in their early stages of development to ensure neat bushy specimens rather than lanky plants devoid of foliage and branches at the base.

You will also learn how to 'harden off' plants correctly, to ensure they do not receive a severe check to growth. This simply means gradually acclimatizing young plants raised in heat to cooler conditions or an outdoor situation.

One of the most testing skills in propagation is looking after the plant material correctly. You may, for instance, find it easy enough to prepare and insert cuttings of various types, but it is the care and attention they receive while they are rooting that generally results in success. Careful attention must be paid to ventilation, temperatures, humidity and control of diseases such as the all-too-common *Botrytis* or grey mould which is the cause of so many dead cuttings. In fact, you must know the optimum conditions required for each particular method of propagation. And timing is also important – if you propagate plants at the correct time of the year then you will be well on the way to successful results.

I hope this has given you some idea of how you will be increasing your horticultural skills. Propagation is not simply taking cuttings or sowing seeds; it demands a knowledge of growing on plants once they have been produced. In the chapters that follow, all the basic methods of propagating plants are covered in detail, and the tables of plants indicate a suitable method or methods of increasing every subject listed. The tables are comprehensive: they list all plants which are commonly grown and many others which are less common.

However, there are bound to be some very uncommon plants which are not mentioned and which some readers will want to propagate. What I would suggest in this instance is that you take a really good look at the plant concerned in order to decide on a suitable method of propagation. If the plant sets a crop of seeds, for instance, then you could try raising it by this means. If plenty of young shoots or stems are produced then it is probably best to try cuttings. Clump formers could well be increased by dividing or splitting the plants into smaller portions. Woody plants like trees and shrubs more often than not respond to the method of propagation known as layering. So do not be afraid to experiment – this adds greatly to the fascination and interest of plant raising. In the past many amateur gardeners have discovered successful methods of increasing plants by nothing more than trial and error. I hope that this adventurous approach continues, for there is always something new to be discovered in plant propagation. Research stations all over the world are constantly finding new or improved methods of propagating plants and much of this information is made available not only to nurserymen but also to amateur gardeners. Likewise amateur gardeners should also share their experiences in plant raising, for instance by chatting to friends, neighbours or garden-club members, or by writing to the popular gardening magazines where their letters could well be published on the 'letters' page'.

4 Raising hardy plants from seeds

Raising plants from seeds is one of the two main methods of plant propagation. (The other method is vegetative propagation which embraces such techniques as cuttings, grafting, budding, division, layering and so on, discussed in Chapters 6–11.) Many of our garden plants set good crops of seeds: they are there for the taking and provide a cheap means of raising new plants. Surprisingly enough many people happily raise summer bedding plants from seeds, yet do not even think about raising hardy plants like trees and shrubs, alpines, bulbs, corms and hardy perennials in the same way. Yet many of these are just as easy to grow from seeds as the popular summer bedding plants.

In this chapter we look in detail at the methods of raising many kinds of hardy garden plants from seeds; Chapter 5 deals with raising tender plants from seeds.

Trees, shrubs and conifers

Trees, shrubs and conifers are expensive to buy from a nurseryman or garden centre and therefore it is far cheaper to raise them from seeds. You can either collect seeds from the plants in the garden, or from friends' and neighbours' gardens, or packets of seeds can be bought from good seedsmen. The catalogues of several leading seedsmen in this country list quite an extensive range of trees, shrubs and conifers.

Collecting seeds However, as a means of obtaining seeds, let us consider collecting from our own plants or from those of friends and neighbours. Also, seeds of many trees can often be collected in the country – very often these are native species but they are none the less attractive for that. Indeed, when it comes to trees, I like to plant some of the smaller native trees in the garden rather than purely exotic species, especially if they are species which are found growing naturally in the locality.

If your area is on chalk then typical indigenous trees may well be beech (*Fagus sylvatica*), ash (*Fraxinus excelsior*), whitebeam (*Sorbus aria*) and yew (*Taxus baccata*). If on the other hand the area is acid heathland then local trees will no doubt include silver birch (*Betula pendula*) and Scots pine (*Pinus sylvestris*). There will also be shrubs indigenous to the area. On chalk there may well be shrubby dogwood (*Cornus sanguinea*), box (*Buxus sempervirens*), guelder rose (*Viburnum opulus*) and buckthorn (*Rhamnus cathartica*), while on acid heathland there will be common gorse (*Ulex europaeus*), common broom (*Cytisus scoparius*) and heather or ling (*Calluna vulgaris*).

There is one important point to remember when collecting seeds – it is only seeds from species of plants (as opposed to hybrids, varieties and cultivars) which will produce plants identical to the parent plants. Species are the true wild plants, unaltered by man, and of course seeds are their natural method of reproduction, so the seedlings will be true to type. If you collect seeds from hybrid plants you will probably end up with a mixed batch of seedlings, with few if any being the same as the parents, so I would generally recommend that you ignore these when it comes to seed collecting and only collect from the true species. If you want seeds of the English oak (*Quercus robur*), for instance, then make quite sure the tree is in fact the English oak and not some other species. There are several good tree and shrub books on the market to aid identification. It is possible to buy packets of mixed hybrids of various shrubs, such as *Buddleia*, from seedsmen.

The seeds of most trees, shrubs and conifers are ready for collecting in the late summer and autumn, although they may be ready any time in spring and summer as well, depending on the particular plant. So keep an eye open for ripening seeds throughout the growing season.

Only collect seeds from vigorous healthy plants as then the vigour will be passed on to the offspring. Diseased plants, including those suffering from the incurable virus diseases, should not be used for seed collecting, as the diseases could be passed on to the seedlings.

Before collecting seeds in quantity do try to make sure that there are in fact seeds in the pods or capsules. Split a few open to check. Also try as far as possible to make sure that the actual seeds are viable, that is, they are alive and well. This can be difficult if not impossible with many of the small seeds and you just have to take a chance that they will germinate when sown. Larger seeds, however, can be tested by cutting open a few and if they are fat and plump inside then there is a good chance that others will germinate. If there is nothing inside the samples, the seeds are not viable.

It is best to collect seeds when they are dry, so try to choose a warm sunny day for the job. It does not matter if fruits and berries are wet when collected, but the 'dry' seeds, which are contained in pods and capsules, should be as dry as possible when collected, as this makes later drying so much easier (see the next section on 'Drying and Cleaning').

The seeds of most trees and shrubs should be collected only when they are ripe, but before the plant starts dispersing them, especially in the case of fine or small seeds. With large seeds, such as acorns, chestnuts and horse chestnuts, it is better to wait until they drop and then pick them up off the ground. The same applies to such trees as maple (*Acer*) and ash.

With most plants it is quite easy to tell when the seeds are ripening. With fruiting and berrying trees and shrubs the fruits and berries generally turn to various colours – red, orange, pink, white, yellow, violet or whatever the normal colour may be. A great many trees and shrubs, however, produce their seeds in dry pods and capsules and these come in many shapes and sizes. The pods and capsules usually begin green, but as the seeds start to ripen they turn brown or maybe greyish or blackish. Some common examples of plants which produce their seeds in pods or capsules include broom (*Cytisus*), *Laburnum*, *Colutea*, *Genista*, *Robinia*, *Rhododendron* and *Hypericum*. When the seeds are ripe, these pods and capsules split open and the seed is ejected, often violently, so that it is spread over a wide area. So they must be collected just before this happens otherwise the seeds, which are often small, will be lost.

Some plants produce what are known as 'winged' seeds – the actual seed has attached to it a papery appendage shaped rather like a wing and this helps in dispersal. Winged seeds generally turn from green to brown before they are released from the parent plant. When they leave the tree the seeds twirl through the air and land some distance away from the parent tree. Maple, sycamore and ash are common examples of plants with winged seeds.

Conifers of most kinds also have winged seeds, like pine (*Pinus*) and larch (*Larix*). With conifers, however, the seeds are enclosed in a cone–

27

a woody seed container made up basically of scales, with a seed beneath each scale. Cones turn from green to brown as the seeds ripen so this is an indication of when to harvest them. Do not wait for cones to fall before collecting as very often they remain on the trees a considerable time after the seeds have been shed. When the seeds are ripe the scales open in order to release them, so if the cones on your trees are open then it is almost certain that the seeds have been dispersed.

Some plants produce plumed or feathery seeds, as in the case of *Clematis*. The plumes turn silvery when the seeds are ripe. Many readers may be familiar with the silvery seed heads of the native *Clematis vitalba*, Traveller's Joy, seen on the hedgerows during the autumn.

There are various methods of collecting seeds. Very large seed containers such as berries and fruits (for instance, rose heps, *Cotoneaster*, *Ilex* (holly) and *Malus* (crab apple) and cones can be picked individually. Seed heads, containing many small capsules or pods, can be gathered complete – for instance, *Rhododendron* produces seed heads, and so does lilac (*Syringa*). Bunches of winged seeds, for instance maple (*Acer*) and ash (*Fraxinus*), can be gathered off the tree with a pair of secateurs. Sometimes, if seeds are difficult to reach, they can be shaken off the tree or shrub onto a sheet of polythene laid on the ground below. A long pole is often useful, too, for knocking down seeds, such as cones, which are high up in the tree. As the seeds are collected they should be placed in separate paper or polythene bags, ensuring that each bag is adequately labelled (type, date of collection, place, etc.).

Drying and cleaning Some seeds will need drying after collection, before they are stored until sowing time. This applies to seeds in dry pods and capsules, the winged and plumed seeds, and also conifer seeds. Drying involves laying out the pods and capsules for several weeks on sheets of newspaper in a warm place. The ideal site is on the bench in a greenhouse; they should be given full sun and have well-ventilated conditions. A windowsill indoors, provided it is sunny for at least part of the day, would also be a suitable place for drying seeds. Under such conditions the seeds will dry and finish the ripening process, and you will find that many of the pods and capsules split open and release the seeds on to the paper. Cones and explosive seed heads should be placed in cardboard boxes or something similar to prevent the seeds from scattering.

If you have collected conifer cones you may find that some of them open satisfactorily under the conditions I have mentioned, but on the other hand the cones of some species are reluctant to open. Cones which are collected commercially in large quantities for forestry purposes are very often put into kilns for several hours and are heated to a high but safe temperature to encourage the cones to open quickly. Of course the amateur gardener cannot use this practice, so for those cones which persistently remain closed, I would suggest that they are kept in the warmest possible place. They could be put in a heated airing cupboard, for example, or they could be laid in a seed box and placed above heating pipes or maybe a radiator – somewhere, in fact, where heat will circulate through and around them. It could take many weeks for some cones to open and shed their seeds, so you may have to be patient.

Once the seeds have dried they will have to be cleaned prior to storage. The seeds must be separated from the pods and capsules: as already mentioned, some may split open during drying and release the seeds, in which case it is an easy matter to gather up the seeds from the newspaper, discarding the pods and capsules.

However, some pods and capsules will not split open and therefore one should gently crush them or rub them in the hands to allow the seeds to drop out. You will now have a mixture of seeds and debris – correctly known as chaff – and the seeds must be separated from this dust and other fragments. There are various ways to do this. If you have a lot of seeds to clean it may be worth while investing in a set of seed or kitchen sieves with various mesh sizes. A batch of seeds and chaff is sifted, perhaps through several sieves, until only very clean seeds are left. (The seeds fall through the mesh, leaving chaff and fragments in the sieves.)

Another method of cleaning seeds is to spread them out on a sheet of newspaper and gently blow away the chaff. You may not remove all the dust by this method, but you will end up with an acceptably clean batch of seeds.

In commercial practice the wings on winged seeds are removed mechanically, but I would advise the amateur not to worry about doing this. They cause no harm whatsoever when left on and do not affect storage or germination. In fact, if wings are not removed very carefully it is possible to damage the seeds and then they may not germinate.

Storing seeds Having cleaned the seeds as much as possible they should then be stored correctly until sowing time in the spring: put them into paper seed packets, envelopes or special linen seed bags, and then seal them with adhesive tape. Be sure to label each packet with its full name and, for reference, the date and place of collection. Seeds can be stored in a cool, dry, airy place but they must not be subjected to frost. They must also be kept free from the ravages of mice, as these creatures like nothing better than a diet of seeds over the winter. If you are storing the seeds where you think mice may get at them, then I would suggest that you place packets in a polythene container with a secure lid – large ice-cream containers are useful – or in a metal biscuit tin. But probably the best place to store seeds is in a cool room indoors, in which case they should not be consumed by mice!

Seeds can also be stored in a refrigerator at a temperature of 1–5°C (35–40°F), but do not allow them to freeze, of course.

A better method for the storage of maple (*Acer*), ash (*Fraxinus*) and beech (*Fagus*) seeds is to mix them with very slightly moist peat and put this in polythene bags, which should then be sealed and stored in a refrigerator, again at a temperature of 1–5°C (35–40°F). This prevents

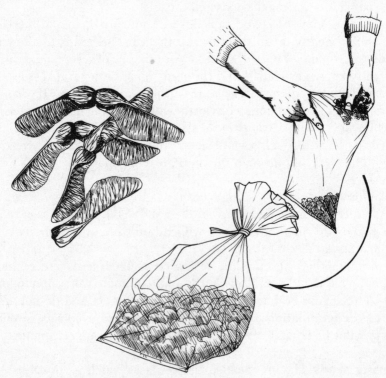

Figure 1 Seeds of maple (*Acer*) can be mixed with slightly moist peat and stored in polythene bags in a refrigerator

the seeds from drying out, which could reduce the percentage germination, and they keep very much better under these conditions. Very often in the spring I find that the seeds are starting to germinate, even though the temperature is low; the seeds must be sown before the young shoots become very long.

Stratification of fruits and berries So far we have only considered the storing of 'dry' seeds, so let us now turn to the fleshy fruits and berries. Examples of the subjects that come in this category include *Cotoneaster*, *Berberis*, holly (*Ilex*), mountain ash and whitebeam (*Sorbus*), crab apple (*Malus*), and ornamental cherry (*Prunus*). These, and many other similar fleshy subjects, are stored by a method known as stratification.

The object of the exercise is to soften the hard seed coats to ensure quicker germination when the seeds are sown. Although the seeds are surrounded by soft flesh, the outer layer of each seed (the seed coat or testa) is extremely hard and unable to take in water (which is necessary for germination to commence) until it has become much softer.

There are two ways of stratifying seeds and I consider the following to be the most successful. The berries or fruits, once collected, should be laid out on a wooden bench or some other suitable surface and macerated to expose the seeds: simply crush them with a block of wood until the seeds are exposed. If you wish to separate the seeds and pulp, then place the whole lot in a bucket of water for several days. Actually this is a good way to test the viability of seeds, for the pulp and the light non-viable seeds will float while the sound, heavy, viable seeds will sink to the bottom of the bucket. At this stage the water and pulp can be drained off, leaving a clean batch of good seeds. But the seeds and pulp can be sown together without detriment.

The seeds, or the mixture of seeds and pulp, should then be mixed with a volume of moist, coarse horticultural sand; use one to three times their own volume of sand and mix thoroughly. Place each type into a clay or plastic flower pot or a tin with holes punched in the base to allow excess water to escape. A 2.5 cm (1 in.) layer of sand should then be placed on top, and the containers labelled with a plastic label. Use a waterproof pen for writing. (See Figure 2.)

The other method of stratifying seeds is to place single layers of uncrushed fruits and berries between 2.5 cm (1 in.) layers of sand in the containers. First put a layer of sand in the bottom of the container, then a layer of berries, followed by a layer of sand, and so on, finishing up with a layer of sand at the top. With this method the fleshy parts of the fruits or berries take a long time to decompose, whereas by macerating them decomposition is very quick, and the seed coats are then more quickly and thoroughly softened.

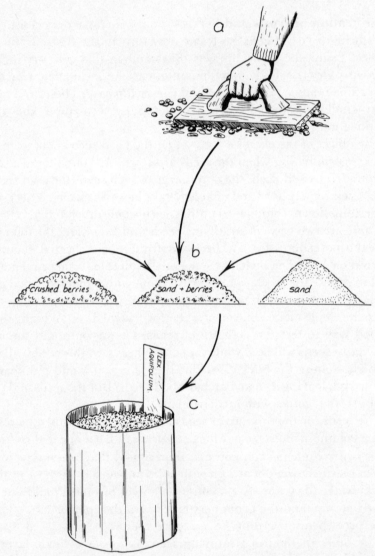

Figure 2 Preparing seeds of holly (*Ilex*) for stratification: (a) The berries are crushed with a wooden presser. (b) The crushed berries are then mixed with moist sand. (c) The mixture is placed in tins or pots and labelled

Once the fruits and berries have been placed in containers, the containers should be stored in the coldest aspect possible out of doors – preferably a north-facing position. There the seeds will be subjected to alternate freezing and thawing and it is this action which softens the seed coats. The stratification process may last from six to eighteen months according to the particular subject. For instance, the following need six months' stratification prior to sowing: *Arbutus*, *Berberis*, *Cotoneaster*, *Euonymus*, *Juniperus*, *Mahonia*, *Malus*, *Prunus*, *Pyracantha*, *Skimmia*, *Sorbus*, *Symphoricarpos*, *Taxus* and *Viburnum*. Subjects which need eighteen months include: *Crataegus*, *Ilex*, *Magnolia* and *Rosa*.

During the stratification period the seeds will need protecting from rodents and birds. The best way is to completely enclose the batch of containers with fine-mesh wire netting; alternatively a piece of wire netting could be laid over the top of the containers and securely held down to prevent rodents from getting underneath.

It is doubtful if the containers will need to be watered during the winter, but at the onset of drier weather in the spring you will need to keep the sand moist by applying water with a rosed watering can.

Once the seeds have been stratified for the required period, the seeds and sand are sown together – it is virtually impossible, and completely unnecessary, to separate the seeds from the sand.

Autumn sowing There are some species which are best not stored at all but sown as soon as they have been collected in the autumn. Included here are oak (*Quercus*), horse chestnut (*Aesculus*) and sweet chestnut (*Castanea*). Seeds of these trees do not store successfully and, in fact, can lose their viability over the winter. The germination process starts as soon as the seeds have been shed from the trees. Another seed which can be autumn sown to ensure really good germination is ash (*Fraxinus*). Indeed, commercially the seeds are often collected when they are still green and sown immediately to ensure really good germination in the spring.

Methods of sowing There are two methods of sowing tree and shrub seeds – in outdoor seed beds or in a heated greenhouse or propagating case. Most species can be sown out of doors, so let us consider this method first.

Seeds can be sown out of doors in March or April, as soon as the soil is in a suitable condition. Ideally the seed beds should be partly prepared the preceding autumn, so that only final preparation of the surface is necessary prior to sowing in the spring. Prepare the site for the seed beds by thoroughly digging in the autumn to the depth of the fork or spade. Ensure that the roots of perennial weeds are removed. If the soil is poorly drained it is a good idea to throw it up into raised beds to ensure good drainage. I generally throw the soil up to form 1 m (3 ft) wide beds about 7.5–10 cm (3–4 in.) high, which can then be left over the winter.

In the spring, just before sowing, carry out final surface preparations. The soil can be lightly forked over and levelled with the fork, and then firmed with the heels by systematically treading it. Then a compound fertilizer (preferably a slow-release type) can be applied according to the manufacturer's instructions. Rake the surface to a fine tilth – this will also help to incorporate the fertilizer in the soil surface.

Now all is ready for sowing. One can either sow broadcast or in drills across the bed – I much prefer the latter method as it makes subsequent weeding so much easier. Use the corner of a draw hoe and take out drills 10 cm (4 in.) apart. The depth will depend on the size of the seeds – the rule is, of course, to sow small seeds less deeply than larger ones – but it generally varies from 0.5–1 cm (¼–½ in.).

After sowing the seeds thinly, they can be covered either with fine soil or preferably with pea shingle or gravel. My preference is for pea shingle because it prevents the surface of the bed from drying out rapidly in hot weather. Also there is no 'capping' of the soil during heavy rain or watering, which can inhibit germination. Shingle also helps to protect the seeds from the ravages of birds.

The seed beds must be kept watered in dry weather as dry conditions can inhibit germination, and it is also vital to pay attention to weeding, which must be done by hand, as weeds can very quickly smother germinating seeds. In very hot weather it pays to erect shading netting over the beds to prevent the seedlings from becoming scorched by the sun.

Some species, especially those with very fine or small seeds like *Rhododendron* and *Erica*, are best sown in containers and germinated in a heated greenhouse or in a propagating case. Species with larger seeds are also sown under glass sometimes, as they generally germinate better with some heat. Those which do well under glass include *Arbutus*, *Calluna*, *Camellia*, *Colutea*, *Cytisus*, *Erica*, *Genista*, *Hamamelis*, *Hypericum*, *Mahonia*, *Pernettya*, *Piptanthus*, *Potentilla*, *Rhododendron*, *Robinia*, *Spartium* and *Viburnum*.

Seeds can be sown in pots of a suitable size, or in seed trays,

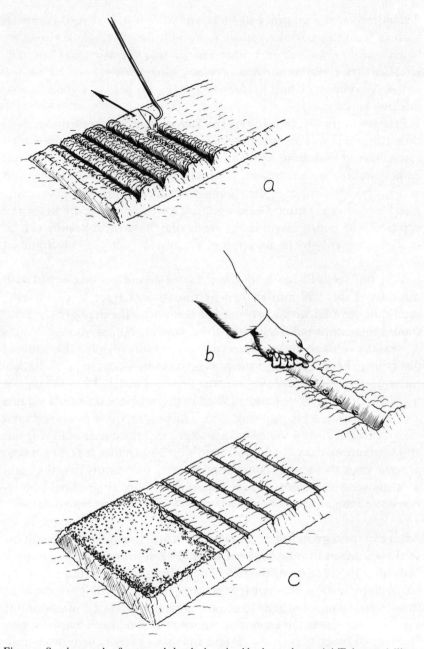

Figure 3 Sowing seeds of trees and shrubs in raised beds outdoors: (a) Take out drills 10 cm (4 in.) apart, the depth depending on the size of seeds. (b) Sow the seeds thinly by hand. (c) The seeds can be covered either with shingle or gravel (as shown here) or with fine soil

depending on the quantity to be sown. A loam-based seed compost, such as John Innes seed compost, is suitable for all subjects. However, when sowing *Ericaceous* and other lime-hating species, like *Erica*, *Calluna*, *Camellia*, *Pernettya* and *Rhododendron*, you will need to use an acid or lime-free compost; I find that moist sphagnum peat is a good sowing medium for these.

Prepare a fine, level sowing surface and remember to sow the seeds as thinly as possible. This is not easy with tiny or dust-like seeds, but they can be mixed with some very fine dry silver sand to bulk them up and make handling easier. Sow the mixture of seeds and sand thinly and evenly over the surface of the compost. A suitable quantity of very fine seed for sowing a standard-size seed tray would cover the diameter and depth of a ½ pence piece! Large seeds like those of *Hamamelis* can be spaced out evenly by hand, approximately 1–1.5 cm (½–¾ in.) apart each way.

Very fine seeds like those of *Rhododendron* should not be covered with compost as this will inhibit their germination. Larger seeds, though, should be covered with a layer of fine compost – the depth of the layer should equal approximately twice the diameter of the seeds.

Next the seeds should be watered. The best way of doing this to avoid disturbing them is to stand the pots or trays in a container of shallow water until the surface of the compost becomes moist. Then the pots or trays should be removed and allowed to drain before being placed in a greenhouse or in a propagating case. The pots or trays can be covered with a sheet of glass and newspaper to prevent the compost drying out and to maintain dark and humid conditions. Turn the glass each day to prevent excessive amounts of condensation from saturating the compost. As soon as germination is apparent the coverings should be removed and maximum light given to the seedlings to prevent etiolation.

Aftercare and growing on of young plants   Let us first deal with the seedlings raised in outdoor seed beds. The seedlings remain in their beds until the autumn or the following spring, at which times they can be carefully lifted with a garden fork and transplanted. They could be transferred to nursery beds to grow to a larger size or be placed in the garden. (For further information about aftercare, see Chapter 13 on Nursery Management.) If you find that some seeds have not germinated, do not destroy the seed beds but leave them intact; in the second spring from sowing you may find that the seeds germinate well.

Now let us look at the requirements of seeds raised under glass. Once the seedlings are large enough to handle easily, but before they become overcrowded in their pots or trays, they should be pricked out, or transferred to other containers to give them more room to develop. Very small slow-growing seedlings can be pricked out into seed trays and when they are starting to fill these they can be moved on into individual small pots, of about 7.5–9 cm (3–3½ in.) diameter. Pricking out into seed trays is discussed in detail in Chapter 5.

More vigorous seedlings and also those which resent root disturbance, such as *Cytisus*, *Genista* and *Spartium*, can be potted direct into 9 cm (3½ in.) pots.

For the majority of subjects John Innes potting compost No. 1 is suitable for pricking off, or an equivalent loamless or soilless mixture. However, for *Ericaceous* and other lime-hating plants you will need to use an acid compost, either loam based or loamless.

After pricking out, water the seedlings and ideally return them to the greenhouse to become established in their pots or trays. Then they should be gradually acclimatized to outdoor conditions. Place them in cold frames and keep the glass covers closed for the first few days; gradually admit more air over a period of several weeks until the lights or covers are open to their fullest extent.

If the seedlings are large enough by the autumn they could be planted out into nursery beds or in the garden. But if they seem rather small and delicate at this time, then it may be safer to overwinter them in cold frames and to plant them out the following spring. Keep the frame lights over them during the winter but ventilate freely on all suitable occasions. Hardy plants do not need coddling – all they need in the seedling stage is protection from the worst of the winter weather.

Alpines

Alpines or rock-garden plants are generally easy enough to raise from seeds but, as with trees and shrubs, I would only recommend that species and not hybrid plants are seed raised. You can either save seeds from your own plants or buy seeds from a seedsman. It is possible to buy mixed hybrids of certain alpines from seedsmen, such as the ever-popular *Aubrieta*. If you are a member of the Alpine Garden Society or the Royal Horticultural Society, then you will no doubt receive seed lists each year which include a range of alpines as well as other kinds of plants. Selecting species from such lists is a good way of obtaining some of the more uncommon alpines. Plant collectors still search the mountainous areas of many parts of the world and collect seeds of alpine plants, and their collections are often included in seed lists too.

Collecting and sowing seeds from your own plants is a useful method of perpetuating certain species, especially those which are impossible to propagate by any other means. Included here are monocarpic species – plants which die after they have set a crop of seeds. Typical examples of monocarpic plants are *Meconopsis integrifolia* (a poppy) and *Ionopsidium acaule*. *Primula viali*, *P. reidii* and *Papaver alpinum*, while not being true monocarpic plants, are nevertheless best raised from seeds each year so that you have a continuous supply of young plants to replace older specimens which are deteriorating in quality.

Some other alpines are impossible to propagate by any method other than seeds, and these include some of the *Androsace* species. Many other alpines set good crops of seeds and are very easily seed raised, so this is as good a method of propagation as any. Examples are given in the tables which appear further on in this book.

Collecting seeds The procedure is similar to that outlined under Trees and Shrubs: the seeds should be collected when they are ripe but just before they are scattered by the plants. This means checking the plants each day, remembering that an indication of seed ripening is a change of colour in the capsules and pods. The seed heads should be collected into paper bags on a dry sunny day and if any pods or capsules split open when collected the seeds will drop into the bottom of the bags.

Drying and cleaning The seed heads are spread on newspaper in a warm, dry sunny place under cover until dry. After two or three weeks the seeds can be removed from the pods and capsules as described under Trees and Shrubs.

Storing seeds Seeds of alpines should be stored in a cool, dry, airy place until sowing time in January or February. Paper seed packets or envelopes should ideally be used and do not forget to label each packet with the full name of the plant plus the date and place of collecting, and seal it. An unheated room indoors is as good a place as any to store seeds of alpines, but do ensure they are not exposed to frost or damp conditions. If stored elsewhere, protect if necessary from mice by placing the packets in a lidded polythene or metal box.

While on the subject of storage, there are some species of alpines whose seeds do not store well as they have a brief life. If stored over the winter you may find that they fail to germinate in the spring simply because they are dead. Seeds of brief viability should therefore be sown as soon as they are collected, which may be in the summer or even the autumn. Often you will find that the seeds germinate very quickly, with up to 100 per cent germination. Some well-known examples of plants with a brief life are *Adonis*, *Androsace*, *Anemone*, *Corydalis*, *Lewisia*, *Meconopsis*, and some of the *Primula* species, especially those in the *Petiolaris* section of the genus, like *Primula edgworthii*, *P. gracillipes*, *P. sonchifolia* and *P. whitei*. I should think twice about buying seeds of those subjects which have a brief life from a seedsman as the germination may not be very good due to the lapse of time between harvesting and sowing.

Methods of sowing Generally I prefer to sow seeds of alpines (apart from those I have just mentioned) as early in the new year as possible – about January or February. Many species require stratification or cold treatment before they will germinate and they will have a chance to receive this if sown early in the year. Alpines which originate from high altitudes certainly need cold treatment prior to germination.

Small quantities of alpine seeds can be sown in 9 cm (3½ in.) pots, but larger quantities can be sown in pans or half pots of a convenient diameter. There is no doubt that clay pots or pans are generally preferred by alpine enthusiasts, but plastic containers can also be used. With clay pots and pans the compost does not remain so wet as in plastic; an overmoist compost is often fatal to seeds of alpines. Ideally each pot or pan should have some drainage material put in the base to ensure the free drainage of surplus water. This practice has largely been discontinued for most subjects in pots, but with alpines I feel it is still well worth while placing a layer of crocks (broken clay flower pots) in the bottom of each seed container; or a layer of shingle could be used instead. Cover the drainage material with a thin layer of peat or leafmould to prevent it from becoming blocked with seed compost.

A suitable compost for alpines is John Innes seed compost. It is essential to have a well-drained compost, and although the John Innes formula ensures this, I still like to add some extra coarse sand to the compost before use to give a really open, well-drained and aerated mix. I should not use a peat-based compost for sowing seeds of alpines, as this can remain too wet after rain or watering. However, it would be

39

suitable for sowing seeds of alpine *Rhododendron* and other *Ericaceous* species, for their compost must be lime free or acid. Even pure peat could be used for these subjects.

As with the sowing of other subjects, alpine seeds must be sown thinly on the surface of the compost to prevent overcrowded and spindly seedlings, which are difficult to prick out. Do not cover very small or fine seeds, like those of alpine *Rhododendron* and *Meconopsis*, but cover larger seeds either with a layer of compost equal to approximately twice the diameter of the seed, or with a layer of chippings or coarse grit. I much prefer the latter covering for alpines, as grit or chippings help to prevent the compost drying out rapidly and preclude the growth of moss and liverwort. It is especially recommended for species which are known to be slow to germinate and which will remain in the pots or pans for perhaps many months.

The compost should be thoroughly moistened after sowing by standing the pots or pans in shallow water until the surface becomes moist. After allowing the containers to drain, they should be placed in the open, preferably in a north-facing aspect, to ensure the seeds receive cold treatment. This will give better germination in the spring once the weather becomes warmer. This treatment, of course, imitates conditions after snow has melted.

Aftercare and growing on of young plants At the onset of warmer weather in the spring the seeds should start to germinate. Species which are known to be slow to germinate could be placed in a heated greenhouse after their cold spell out of doors, as very often this encourages germination.

While on the subject of slow-germinating species, you may find that some do not germinate in the spring following sowing, even after being given a spell in a heated greenhouse. Do not be tempted to throw the seeds away, but stand the containers in a northerly aspect again and keep them watered as necessary. You may then find that the seeds germinate in the second spring after sowing. You will, however, find that the majority of species will germinate without any trouble at all in the first spring.

Once the seedlings are large enough to handle they should be pricked out individually into 7.5 cm (3 in.) pots. A suitable compost is John Innes potting compost No. 1 to which some extra coarse sand or grit has been added to ensure really open conditions and good drainage. To a given volume of JIP1 I generally add about one-third extra sand or

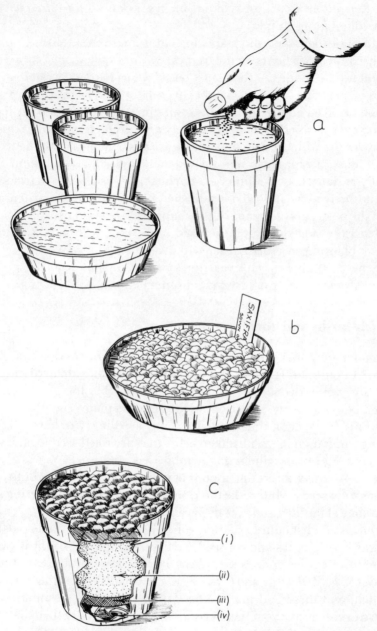

Figure 4 Sowing seeds of alpines: (a) Seeds can be sown in pots, half pots or in pans. (b) They can be covered with a layer of chippings or coarse grit. (c) A completed pot showing materials used: (i) layer of chippings or grit over the seeds; (ii) compost, such as John Innes seed compost; (iii) layer of peat or leafmould; (iv) layer of crocks

grit. Remember to use an acid or lime-free compost for *Ericaceous* and other lime-hating alpines.

After potting, stand the plants in cold frames to establish. You will not need the glass lights on the frames, except in the winter, as alpines should not be coddled – they must be grown as hard as possible. Even though I have recommended lights over them in the winter, the plants should be given ample ventilation at all times by blocking up the lights with blocks of wood. The lights merely give protection from excessive rain over the winter, as young alpines are often very susceptible to wet conditions, causing them to rot. Cold and frost will not kill them. The woolly or hairy leaved alpines are particularly susceptible to excessive wet in the winter, and even established plants on a rock garden may need the protection of a pane of glass supported over them in the winter.

Generally speaking, alpines raised from seeds will be of a suitable size for planting on a rock garden within about a year from pricking out. The most suitable time of year for planting alpines is in the spring, although some people prefer early autumn planting.

Hardy bulbs and corms

I wonder how many people think of raising bulbs and corms from seeds? I doubt if it is a very common practice among amateur gardeners, and yet I do not see why it should not be, for many species germinate extremely well from seed. Indeed, many sow themselves naturally in gardens, and no doubt many seedlings are hoed off soon after germination in the mistaken belief they are weed seedlings. I must confess some bulb seedlings do resemble weed grasses.

It does require a certain amount of patience when raising bulbs and corms from seeds. Many take three or four years to reach flowering size, and some of the lilies can take five years or more.

Once again I should stress that you only raise species from seeds and not the hybrid bulbs and corms. I particularly like raising the dwarf or miniature species from seeds. *Narcissus*, *Chionodoxa*, *Muscari*, *Crocus*, *Tulipa*, *Scilla*, *Galanthus* and *Cyclamen* are so useful for planting on a rock garden, and these and many other bulbs and corms germinate well. You can either save your own seeds or buy from a good seedsman.

Collecting seeds The seeds of bulbs and corms are carried in pods, and as the seeds ripen the pods turn brown and start to split open. For further details of collecting refer to the section on Alpines.

Drying and cleaning The seeds are cleaned and dried by the method recommended for alpines, and full details of the procedures can be found in the Tree and Shrub section of this chapter.

Storing seeds Seeds of bulbs and corms need the same storage conditions as alpines and therefore readers should refer to the section on Alpines.

Methods of sowing Although I have said that the seeds of bulbs and corms can be stored, you will achieve better germination if they are sown as soon as they have been collected from the plants. This may be any time in the summer or autumn, as soon as the seeds are ripe. Or seeds may be sown at the same time as alpines – in January or February.

The sowing procedure described for alpines is also suitable for bulbs and corms, so readers should refer to that section for further details.

Aftercare and growing on of young plants The pots or pans of seeds should be placed in a northerly aspect out of doors so that they are subjected to cold conditions prior to germination. As with some alpines, the seeds of certain species of bulbs and corms may not germinate in the first spring, so keep the containers for another year as germination may be delayed. Some *Lilium* species are slow to germinate, but many other bulbs and corms germinate freely in the first spring after sowing.

It is common practice to leave the seedlings in their pots or pans for a year, or maybe two years, after germination, to allow them to grow to a convenient size for planting out. Seedlings of some subjects are very small and difficult to handle, so it is very important to sow the seeds thinly to ensure that the seedlings are not overcrowded in their containers. All the time the seedlings are in their pots they should of course be watered as required. When the foliage has died down, however, water can be withheld, resuming in late summer or autumn when the tiny bulbs should be making new growth.

Young bulbs must be planted out when they are dormant – which is late summer or autumn for most subjects. Knock them out of their containers and carefully separate them. Immature bulbs are generally planted more shallowly than full-size flowering bulbs. They could be grown on in nursery rows until they reach flowering size, at which stage they can be set out in their final positions in the garden.

Hardy perennials and grasses

Hardy perennials are non-woody plants and the majority of them are herbaceous in habit, that is, the stems die down to ground level in the autumn and new shoots are produced again in the spring. Some perennials, however, are evergreen, such as the hellebores, some *Iris* species, *Bergenia* and also some of the ornamental grasses which are so popular these days for flower arranging. Hardy perennials and grasses are very useful for providing colour and interest from spring to autumn; some even flower in the winter. Herbaceous plants were once grown in their own special borders or beds, and while this method is still used by some of the larger private and public gardens, most amateurs nowadays seem to grow them in a mixed border with other plants, such as shrubs, trees and bulbs.

The main method of propagation of hardy perennials and grasses is division of the plants while they are dormant, and this method is fully described in Chapter 10. However, some perennials cannot be divided as they only produce a single crown, and not a clump of crowns suitable for division, and so raising them from seeds or cuttings is the usual method. Of course, seeds are a good way to raise large quantities of plants quickly and cheaply; any of the species could if desired be propagated by this method. I again emphasize species, because the hybrid plants (and there are many of them among hardy perennials) will not breed true to type from seeds. If the dead flower heads are not cut off, the species will generally produce a good quantity of seeds each year, which can be collected and sown at the appropriate time. Of course, seeds of hardy perennials and grasses can be bought from a good seedsman – both genuine species and also some very attractive mixed hybrids of such subjects as *Lupinus*, *Aquilegia*, *Gaillardia*, *Rudbeckia* and *Delphinium*. One or two of the bigger seed firms in this country have in recent years bred some superb and colourful strains of many of our popular herbaceous plants.

Collecting, drying and storing seeds These aspects have already been adequately covered in this chapter and readers should therefore refer to the appropriate sections under Trees and Shrubs, and Alpines, for further details of collecting. The seeds of the majority of hardy perennials are contained in pods and capsules of various shapes and sizes, while those of grasses are produced in papery protective coverings. Seeds of hardy perennials should be dried after collecting and stored in dry airy conditions until sowing time, as described earlier in this chapter.

Methods of sowing Seeds of hardy perennials and grasses can be sown in the open ground during May or June. The seed bed should be sited in an open, sunny, well-drained position with a reasonably fertile soil, and should be prepared first by digging the site to the depth of a fork and breaking the soil down finely. Remove the roots of any perennial weeds during this initial cultivation. The soil should be firmed by treading systematically with the heels. Now a general-purpose fertilizer can be applied, according to the manufacturer's directions, and this should be thoroughly raked or pricked into the soil surface. Final preparation involves raking the soil surface to produce a fine tilth and a level surface – this is best carried out when the soil is slightly moist, not too wet or very dry.

We are now ready for sowing the seeds. I prefer to sow in shallow drills taken out with the corner of a draw hoe. Seedlings in straight rows are easier to weed and cultivate than those which have been sown broadcast. The seeds should be sown as thinly as possible in the drills and then covered with soil – the depth being up to twice the diameter of the seeds.

After the seeds have been sown the beds must be kept moist, watering if the surface starts to become dry. This will ensure good and even germination. If the seed beds become dry then patchy germination will be the result – or maybe it will be a complete failure. Even after the seeds have germinated, the beds should still be kept moist as tiny seedlings are very vulnerable to drying and will soon die if not watered.

Some hardy perennials are probably best sown in seed trays and germinated in a greenhouse or cold frames. This would certainly apply to those plants with very fine or small seeds as then there is better control over sowing. Seeds which are generally sown in frames or the greenhouse include *Delphinium*, *Lupinus*, *Primula* and *Viola*. They can be sown in seed trays of John Innes seed compost: if the seeds are very fine do not cover with compost; otherwise cover lightly with fine sifted compost.

Aftercare and growing on of young plants Let us first deal with those raised in outdoor seed beds. When the seedlings are large enough to handle easily they should be transplanted into nursery beds to give them more room for development.

Again choose an open, sunny, well-drained fertile site for growing the young plants. Dig it thoroughly and firm well with the heels. Do not forget to remove any perennial weeds. A general-purpose fertilizer can then be applied according to the manufacturer's instructions and this should be well worked into the soil surface.

The seedlings should be carefully lifted with a fork, avoiding as much root damage as possible. It is easier to lift if the soil is nicely moist, but not over-wet. On no account let the roots of the seedlings become dry between lifting and replanting. They could be placed in a bucket of water to keep them moist while planting; this is especially necessary if the weather is warm and sunny.

Line out the seedlings in nursery rows 30 cm (12 in.) apart with 15–20 cm (6–8 in.) between the plants in the rows. Plant with a hand trowel and ensure the roots are well down in the soil and not cramped in the holes. Firm in thoroughly with the fingers. After planting, the young plants should be kept moist to ensure they make steady growth and they must also be hand weeded or hoed regularly to eliminate competition from weeds.

By the following autumn – September or October – the young plants will be ready for setting out in their final positions in the garden. Most should then flower the following year. You may even get a few flowers from some perennials in the first summer or autumn.

Now to those sown in cold frames or the greenhouse. Once the seeds have germinated the seedlings should be pricked out into seed trays to give them more room to grow. Use John Innes potting compost No. 1 or an equivalent loamless type, and prick out the seedlings – forty-eight per tray. Further details of pricking out can be found in Chapter 5. They should then be returned to the cold frames and gradually hardened ready for planting in nursery beds – when they start to fill the trays but before they become overcrowded.

Hardy ferns

The growing of hardy ferns, generally in a 'fernery', with cool, moist, shaded conditions, was very popular in the Victorian era but subsequently these plants lost much of their appeal. However, in recent years interest in hardy ferns has been revived and the plants are now much publicized in books and magazines. There are one or two very good specialist fern books available.

Ferns can be propagated vegetatively by division (see Chapter 10), but another method is from spores, the reproductive bodies found on the undersides of the fronds or leaves, and the ferns' equivalent to seeds. Not all hardy ferns produce spores, but many species do produce them freely under cultivation. They should be collected as soon as they are ripe, and sown under glass. Use the same technique as for greenhouse ferns – the method is fully described in Chapter 5. The young ferns are eventually potted and hardened in cold frames before being planted in the garden.

Hardy annuals

Annuals are plants which grow to maturity, flower and set their seeds within one year. After producing seeds they die. The hardy annuals are very colourful plants for summer flowering and they are often grown in their own special beds or borders to provide continuous colour and interest from late spring until the autumn. They are an inexpensive means of providing colour in a garden, for packets of seeds are still relatively cheap to buy from a seedsman or garden centre. The seeds are sown out of doors in the spring, where the plants are to flower, so no artificial heat is needed to raise them.

In a good seed catalogue there will be many dozens of hardy annuals listed and illustrated. Some packets will be mixed hybrids in a range of different colours, while others will be true species, from which you will be able to save your own seeds instead of buying each year. Details of collecting, drying and storage of seeds will be found under Alpines in this chapter.

Choosing a site Many of the hardy annuals come from hot climates, such as South Africa and Mexico, and so they should be grown in this country in a position which receives full sun and light. A shaded site will result in weak growth and poor flower production. The soil must be well drained, as annuals dislike very wet conditions, and they should not be given a very rich soil as they will then make excessive vegetative growth at the expense of flower production. Some can in fact be grown in very poor soils, such as the ever-popular nasturtium (*Tropaeolum*); these will flower very freely if given a lean diet. Generally speaking, however, a moderately fertile soil is to be recommended for balanced growth and flowering.

47

Many annuals are very suitable for cutting for flower arranging and a good catalogue will indicate which ones are best for this purpose. I generally prefer to grow annuals for cutting in a utility part of the garden, such as near the vegetable plot, as then the main flower display in the ornamental part is not ruined. While on the subject of cut flowers, remember that there is a good range of annual ornamental grasses available these days, which can be cut and dried for winter decoration. There are also 'everlasting' flowers of various kinds which again are excellent dried for the winter.

Some of the dwarf annuals are excellent for filling gaps on a rock garden – again choose a sunny position for them and sow *in situ*. Dwarf annuals can also be grown in the gaps in crazy paving – a favourite technique in cottage gardens.

Preparation for sowing The site for sowing hardy annuals should be thoroughly dug to the depth of a fork or spade in the autumn, removing any perennial weeds at the same time. If the soil is poorly drained, then it would be advisable to dig in a good quantity of coarse sand or grit to help open up the soil and improve the drainage. Garden compost will also help in this respect but only use well-rotted compost; alternatively leafmould or peat could be used. Bulky organic matter like compost, leafmould or peat is especially necessary for very light poor soils which are unable to hold much moisture and therefore dry rapidly in warm weather. Organic matter acts as a sponge in this case and holds a certain amount of moisture which the plants can absorb as required. After autumn digging the soil should be left in this rough state until the spring, so that the elements can work on it; then it will be an easy matter to prepare the surface for sowing in the spring.

Seven to ten days before sowing, a general-purpose fertilizer could be applied to the soil surface, according to the manufacturer's instructions, and pricked into the soil. Just before sowing the final soil preparations can be undertaken. Choose a day when the soil surface is only slightly moist – not too wet or very dry. First break down the soil surface with a fork, and then tread it systematically with your heels to firm it. Now, using an iron rake, rake the surface in various directions to create a really fine tilth and a level surface. All is now ready for sowing the seeds.

Methods of sowing Hardy annuals are sown in the spring as soon as the soil is in a workable condition. Try to sow in March or April, but if you sow too early the soil will still be cold and wet and the seeds may rot instead of germinating. It is far better to wait until soil conditions are really suitable.

If you are sowing for display purposes, such as in a bed or border, then sow each kind in a bold informal group. For impact, each group should ideally be at least 1 m (3 ft 3 in.) by 1 m. Really such a bed or border should be carefully planned beforehand to ensure suitable colour combinations: the bed or border can be marked out before sowing, according to the plan, with the outline of each group defined or marked on the soil with some dry sand. Heights of plants must also be taken into account when planning the scheme so that each group can be easily seen. There is plenty of information available on planning annual borders and readers should therefore refer to appropriate books.

The seeds can then be sown in drills in each group. Take the drills out with the corner of a draw hoe and space them 15 cm (6 in.) apart. Make sure the drills are shallow, as the seeds must not be covered too deeply – a covering of soil equal to twice the diameter of the seed is a general guide. I much prefer to sow each annual in drills rather than broadcasting the seeds, as it makes subsequent cultivations, like thinning and weeding, so much easier. As the annuals grow the straight lines will gradually disappear – the plants will close up and form a solid patch of colour.

The seeds must be sown as thinly as possible in the drills to minimize subsequent thinning out of seedlings, which is both time consuming and wasteful of seeds. After sowing each group use a rake to draw soil over the seeds. I do not generally firm the soil as this can result in a capped surface, especially if water is applied, and this may inhibit germination.

If you are sowing annuals for cutting, first take out straight drills, using a garden line to keep the rows straight. Then sow thinly and cover the seeds as described above.

Some of the very hardy annuals can be sown in the autumn, in September or early October, to give earlier flowering the following year; examples will be found in the tables. However, autumn sowing is not to be recommended if your soil is heavy and lies wet over the winter because the seedlings will simply rot. A light, sandy, well-drained soil is more suitable for this purpose. Actually the annuals could be covered with cloches over the winter to protect them from excessively wet conditions.

Sweet peas (*Lathyrus*) are very often sown in September or October to give early blooms the following year. They should be sown at five to seven seeds per 9 cm (3½ in.) pot of John Innes potting compost No. 1, then germinated and overwintered in well-ventilated cold frames. Do not coddle the young plants; they must be grown as hard as possible. The seedlings should be kept just moist over the winter and then they can be planted in their flowering positions in the following April. Sweet peas may, of course, be sown direct in their flowering positions in March or April, but this means they will flower later.

Aftercare of annuals After sowing, water the seeds in if the soil is dry, either with a rosed watering can or with a garden sprinkler. Thereafter water whenever required throughout the spring and summer.

Another important operation is weed control: this must receive regular attention either by hoeing or by hand weeding. If you have sown in rows as suggested, rather than broadcast, then it will be an easy matter to hoe between the rows with, preferably, an onion hoe. Eventually the plants should close up and suppress any further weed growth.

As soon as the seedlings are large enough to handle they should be thinned, if necessary, to stand 15 cm (6 in.) apart in the rows. Either pull out the surplus seedlings with your fingers, or use an onion hoe to chop them out. When you have completed thinning it is a good idea to water the remaining seedlings to ensure they are still well settled in the soil— the soil may have become loose around them during thinning.

Some annuals are dwarf in habit, but others may grow to about 1 m (3 ft 3 in.) in height; these taller kinds usually need some means of support as their stems are generally rather weak, and the plants could be flattened during high winds or heavy rain. I find the best way of supporting annuals is to insert twiggy hazel sticks ('pea sticks') around and between them before they become too tall. The plants will then grow up through these sticks and completely hide them, and at the same time they will be adequately supported. Be sure not to have the sticks taller than the eventual height of the plants.

By autumn the plants will be past their best: as soon as the display is over the plants should be pulled up and put on the compost heap— after collecting any seeds of species which you wish to save, of course!

Hardy biennials

A biennial plant is one which takes two growing seasons to come into flower and set its seeds, after which it dies. In gardening we use the term rather loosely – for the sake of convenience we also include a number of short-lived perennials which we grow as biennials: that is, we pull them up and throw them away after they have flowered. Although they could be kept for a few years, they start to deteriorate in quality after the first flower display.

The hardy biennials (and I am including short-lived perennials here as well) are extremely useful plants for colour either in the spring or summer. Many of the spring-flowering kinds are used for bedding displays, perhaps in conjunction with spring bulbs like tulips and hyacinths. Examples are forget-me-nots (*Myosotis*), wallflowers (*Cheiranthus*), polyanthus (*Primula*) and double-flowered daisies (*Bellis*). The last three subjects are in fact short-lived perennials. Some of the early summer flowering biennials are excellent border plants and can often be used as cut flowers. Many are old-fashioned cottage-garden plants, like sweet williams (*Dianthus*), Canterbury bells (*Campanula*) and foxgloves (*Digitalis*). Honesty (*Lunaria*) is another old favourite, and this flowers in the spring. You will find that some of these subjects sow themselves freely, like foxgloves, honesty and forget-me-nots.

Seeds of all these subjects can be obtained from seedsmen. Mixed hybrids and strains are available, and generally packets of single cultivars can also be obtained if desired. There is a big range of named cultivars of wallflowers, to take just one example.

You could of course save the seeds of hybrid plants but there is no telling what the resultant seedlings will be like. It is I must admit rather fun saving your own seeds of sweet williams and wallflowers, for one often gets a good mixture of colours in the seedlings, and some exciting and unusual colour combinations.

The biennials are generally sown out of doors in May or June to produce plants for flowering in the following year. To obtain large plants of polyanthus, however, sow the seeds much earlier in the year – between January and March – as polyanthus are rather slow growers.

Site preparation and sowing The seed bed is prepared as described under Hardy Perennials and the method of sowing is also the same, so readers should refer to that section. I would suggest, however, that the seeds of polyanthus are sown in seed trays and germinated in a greenhouse or cold frame, for they are very small and need to be sown on a fine level surface of John Innes seed compost. As the seeds are so small they should not be covered with a layer of compost. Polyanthus like

51

moist soil conditions and so the seed trays should be kept moist at all times. Provide shade during periods of hot sunshine to prevent scorching of the seedlings and to prevent rapid drying of the compost.

Aftercare and growing on of young plants When the seedlings of biennials are large enough to handle they should be transplanted to nursery beds to grow into larger plants, ready for planting in beds and borders during October and November.

Choose an open sunny position for the beds, except for polyanthus which prefer moist, cool, semi-shaded conditions. The beds should be prepared as discussed under Hardy Perennials, and the planting technique is also the same. The nursery rows should be spaced 30 cm (12 in.) apart with 15 cm (6 in.) between the plants in the rows.

With wallflowers it is recommended that the tap-root of each plant be broken off to encourage a fibrous root system to develop. Young plants with a large fibrous root system transplant far more successfully in the autumn than plants with a single tap-root and few fibrous roots.

Keep the young plants well watered and weeded throughout the growing season and then you will have strong healthy plants for use in the autumn.

Once polyanthus seeds have germinated in the trays the seedlings should, when large enough to handle easily, be pricked out into seed trays to grow to a larger size; then they can be planted in cool, moist nursery beds. Use John Innes potting compost No. 1 for pricking out; and space the seedlings at forty-eight per tray. They can be returned to cold frames to grow, keeping them moist and shaded from strong sunshine. Gradually harden the seedlings before planting in nursery beds.

Actually I have pricked out polyanthus seedlings from trays direct into nursery beds, but this is a fiddly process and I much prefer to prick out into trays and then plant larger specimens. However, direct pricking out does dispense with one operation and the plants, of course, suffer less check to growth.

5 Raising tender plants from seeds

A wide range of greenhouse plants and house plants, including flowering pot plants, tender perennials and shrubby types as well as subjects like cacti and succulents, can easily be raised from seeds provided you can supply sufficient heat. Many gardeners also obtain pleasure from sowing pips and stones of fresh tropical fruits. In a heated greenhouse you can also raise your own summer bedding plants, both the half-hardy annual types and the tender perennials, thus saving a great deal of money each year on boxes of bedding plants.

You may be able to save your own seeds from some subjects which set good crops, but again I would advise collecting only from true species, not from hybrid plants and cultivars which will not breed true to type. Flowers under glass or indoors may need hand pollinating to ensure that seeds are produced. If you look through a good seed catalogue you will find a wide range of tender plants on offer, from greenhouse shrubs and pot plants to summer bedding plants. Packets of seeds are still very good value for money and very much cheaper than buying plants.

The raising of greenhouse ferns from spores is also covered in this chapter.

Greenhouse and house plants

Sowing the seeds The main sowing period for greenhouse and house plants is January to April, although some subjects should be sown at other times of year in order to obtain flowering in the appropriate season. You may also wish to sow certain pot plants at intervals in order to have a long succession of flowers. So it will be necessary for readers to refer to the tables of greenhouse and other tender plants at the end of the book to ensure sowing is carried out at the best possible time of year.

Suitable containers for sowing reasonable quantities of seeds are standard-sized seed trays in wood or plastic. I would recommend that you choose shallow trays for seed sowing as then you will need far less compost. Seeds do not require a great depth of compost and trays about 2.5–3.5 cm (1–1½ in.) in depth are perfectly adequate. If you only have very small quantities of seeds then you could sow in pans or half pots of appropriate diameter. These are sometimes available in clay, but more often than not nowadays they are made of plastic.

The usual compost for seed sowing is the loam-based John Innes seed compost. But there are various soilless seed-sowing composts available now which are composed of peat with fertilizers added: these generally give exceedingly good results.

The compost should be made moderately firm in the containers by pressing all over with the fingers, paying particular attention to the corners and sides of the trays or pans, and the surface should be

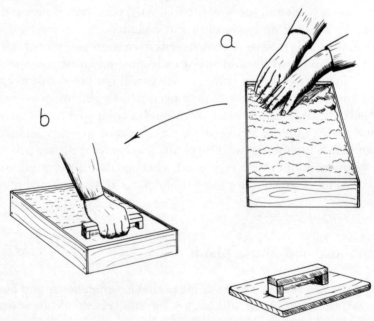

Figure 5 Preparing a seed tray for sowing: (a) The compost should be firmed all over with the fingers, paying particular attention to the corners. (b) The surface should be made smooth and level by pressing with a flat piece of wood or a home-made wooden presser

perfectly smooth and level: press it with a flat piece of wood. If you intend sowing very fine dust-like seeds then you could use a small-mesh sieve to sift a layer of very fine compost over the surface before finally firming it. The final level of the compost should be about 6–12 mm (¼–½ in.) below the top of the container, to allow room for subsequent watering.

Seeds should be sown thinly and evenly; avoid dropping the seeds in thick patches and leaving other patches devoid of seeds. The best way to ensure even sowing of very fine dust-like seeds is to mix them with a quantity of dry, very fine silver sand to make handling easier.

There are lots of methods of ensuring thin even sowing but I can thoroughly recommend the following procedure. In the palm of one hand hold a sufficient quantity of seeds for the particular container. Then raise this hand 10–15 cm (4–6 in.) above the surface of the compost. Now move your hand to and fro across the surface of the compost at the same time gently tapping it with your other hand to release the seeds slowly – I must emphasize the *slow* release of the seeds to avoid thick patches – and you should find that the seeds scatter evenly over the surface of the compost. Sow half the quantity of seed in one direction, say from one end of a seed tray to the other, and then sow the remaining half in the other direction, say from one side of the tray to the other. With a little practice you should find that this method gives a very even sowing. If the very small seeds stick to the palm of your hand then you could place them on a sheet of paper and again sow in the way I have described, but I find that it is not as easy to control the trickle of seeds from a sheet of paper. (See Figure 6.)

The next stage is to cover the seeds with a layer of compost, if they require it. Some seeds should not be covered – this applies to very fine dust-like seeds such as those of *Begonia*, for a layer of compost over these would prevent germination. But other seeds should be covered with a layer of compost equal to twice the diameter of the seeds. The easiest way is to sift a layer of compost over them, using a fine or medium sieve, and making sure that the compost is of a uniform depth all over the seeds, otherwise germination will be uneven.

Very large seeds and the modern pelleted seeds can, of course, be hand spaced evenly on the surface of the compost.

Once the seeds have been sown they will need watering, but I would not recommend overhead watering with a rosed watering can as this is liable to disturb the seeds, especially the dust-like ones which do not have a covering of compost. Instead, stand the containers in shallow

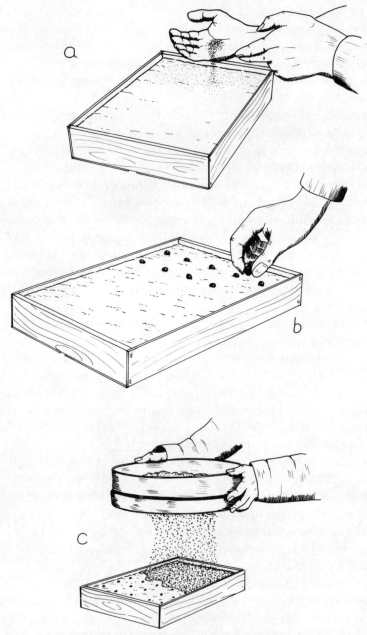

Figure 6 Sowing seeds in a tray: (a) To ensure thin even sowing the seeds can be sown from the palm of one hand. Gently tap the hand to slowly release the seeds. (b) Large seeds can be spaced out by hand. (c) Seeds can be covered by sifting a layer of compost over them, using a fine or medium sieve

trays of water and leave them until the surface of the compost becomes moist. Then they should be removed and allowed to drain before being placed in germinating conditions. Instead of using plain water for moistening the compost you could use a solution of a preparation formulated to prevent damping-off diseases in seedlings. (Damping off is very common in seedlings raised under glass – the seedlings suddenly collapse at soil level and die. So it is worth guarding against this trouble right at the very start.) If you go into a gardening shop or garden centre you should find a suitable product.

The best place to germinate the seeds is in an electrically heated propagating case, with a temperature of 18–21°C (65–70°F), or you may have a bench with soil-warming cables on it and this would also be suitable. Set the thermostat at the temperature range indicated above. If you do not possess either, stand the containers on the greenhouse bench and provide as high an air temperature as possible. It is obviously much more expensive to heat the whole greenhouse to 18–21°C so in the long run it may be better to invest in an electrically heated propagating case, or soil-warming cables for the bench. Many people germinate seeds in a warm place indoors, or on a warm windowsill, but once the seeds have germinated try to ensure that they receive as much light as possible otherwise the seedlings will become drawn and spindly.

To ensure the compost does not dry out before germination occurs the containers can be covered with sheets of newspaper, brown paper or even black polythene. Such a covering also helps to maintain an even temperature over the surface of the compost. As soon as the seeds germinate the covering material must be removed as from now on the seedlings must receive maximum light. If given insufficient light they will become drawn or etiolated, that is, they will have thin weak stems and will probably not develop into good-quality plants. As soon as germination occurs, therefore, place the containers on the greenhouse bench to ensure the seedlings receive really good light.

When the seedlings are large enough to handle easily they should be pricked out or transplanted into other containers to give them room to develop. If they are allowed to become overcrowded in their sowing containers they will again become drawn or etiolated. Generally standard plastic or wooden seed trays are used for pricking out, inserting forty to fifty-four seedlings per tray. Use deep seed trays this time – about 5 cm (2 in.) in depth. If you have only a small quantity of seedlings, or are growing such things as flowering pot plants, the

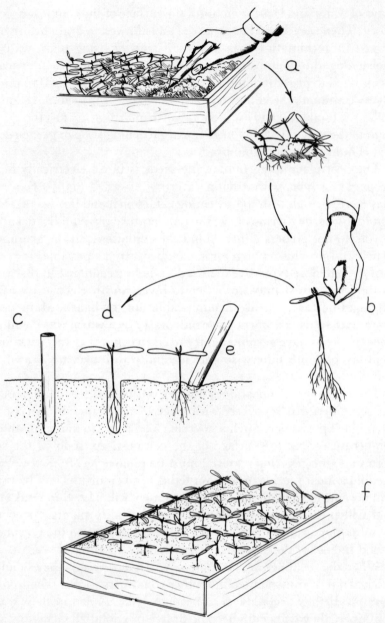

Figure 7 Pricking out seedlings: (a) Carefully lift seedlings with the aid of a dibber and separate them. (b) Always handle seedlings by the cotyledons or seed leaves. (c) Make a hole for each seedling with a wooden dibber. (d) The roots of the seedling should dangle straight down in the hole. (e) Gently firm in the seedlings with the dibber. (f) Seedlings should be evenly spaced out in rows across the tray

seedlings could be pricked out into individual small pots, say of 6 cm (2½ in.) diameter. If you prick out into seed trays first, then you will have to pot the young plants when they have filled the trays. It is really up to you whether you use trays or small pots for pricking out.

A suitable compost for pricking out is John Innes potting compost No. 1, or an equivalent loamless type.

Seedlings should always be handled by the seed leaves or cotyledons (the first leaves produced); never hold them by the stems as these will be very soft at this stage and therefore easily damaged. Seedlings should be carefully lifted out of the seed trays with a dibber (a small length of wood shaped rather like a blunt pencil). Insert this under the roots and lift out a few seedlings at a time.

The boxes or pots should first be filled with compost and then a hole for each seedling made with the dibber. Ensure that the hole is deep enough to allow the roots of the seedlings to dangle straight down – roots should never be restricted in any way and must certainly not be allowed to turn upwards as this will retard the growth of the seedlings. Each seedling should be inserted almost up to its cotyledons and then very gently firmed in with the dibber. There is no need for excessive firming as subsequent watering will settle the compost around the roots, and in any case too much firming would damage the tender roots and stems.

After pricking out thoroughly water the seedlings with a fine-rosed watering can. Again you could use a solution of a fungicide formulated to prevent damping-off diseases, as even after pricking out the seedlings are still susceptible to this trouble.

The seedlings are now grown on, in the greenhouse or indoors in conditions suited to the particular plants. It may be advisable to shade seedlings from strong sunshine to prevent leaf scorch.

Half-hardy annuals and perennials

Half-hardy annuals and perennials, or summer bedding plants as they are more commonly called, are easily raised from seeds in a heated greenhouse. The seeds are sown, germinated and the seedlings pricked out as described under Greenhouse and House Plants. There are some minor differences, for instance bedding plants are generally pricked out into seed trays at forty to fifty-four seedlings per tray.

The sowing period for summer bedding plants is from January to March or early April. You will need plants for bedding by the end of May or early June – when they are just coming into flower – so it is important to sow at the right time. Some plants are slow growers and therefore need sowing in January or February to ensure flowering-size plants by late May. Others are quicker and can be sown in February or early March, while yet others are very fast developers and can be sown any time in March or even into early April. You will need to refer to the tables at the end of the book to find out the best sowing times for particular subjects.

Once seedlings have been pricked out they should be grown in a heated greenhouse – certainly in one which is frost free. Ideally the seedlings should be given warmth, adequate light and moisture and plenty of ventilation during the day if the weather is suitable, that is, not frosty or foggy.

Several weeks prior to planting in the garden the young bedding plants should be properly hardened in a cold frame. This procedure is fully explained in Chapter 13, which deals with Nursery Management.

Greenhouse ferns

Some greenhouse ferns can be raised from the spores which are found on the backs of the fronds. Spores are the reproductive bodies and are the ferns' equivalent to seeds. Many species are difficult, or even impossible, to raise from spores, while others may not even produce spores under cultivation. However, those ferns which are easily increased by this method include most of the *Adiantum* species and many of the *Pteris* species.

The spores should be collected when they are ripe so the backs of the fronds must be inspected frequently. As soon as the sori or spore cases start to turn brown the fronds should be cut and dried in paper bags which should be tied at the top. Once the spores have dropped off the fronds into the bottom of the bag they should be sown. Fresh spores germinate very much better than spores which have been stored for a time.

Use sterilized or new pots or pans and place a layer of drainage material, such as shingle, in the bottom. A suitable compost is fine loam to which has been added finely crushed brick or coarse grit. The loam should be sterilized before use; alternatively peat could be used as this is more or less sterile. The surface of the compost should be made really fine and even, for the spores are minute. Then water the containers and allow to drain before sowing. Now sow the spores very thinly and evenly over the surface and on no account cover with soil.

Cover the containers with glass and stand them in shallow water in a close heated propagating case. Shade from sunshine but remove shading during dull weather.

In these conditions the spores should germinate and form minute plants (correctly called prothalli). These should be pricked out into pots or trays of the same soil. Prick out very small patches, spacing them about 2.5 cm (1 in.) apart – they can be handled with a small notched stick – and simply press the patches lightly into the surface of the compost.

After pricking out keep the small plants in warm moist conditions until they are well established. Again, a heated propagating case would be ideal. The pots or trays can again be stood in shallow water as it is not advisable to water the young plants overhead until fronds have been formed.

The young plants will eventually need to be potted into individual small pots. This is still quite a delicate operation. They can then be grown on in the normal greenhouse atmosphere. The young ferns should be potted into slightly larger pots before their first pots become full of roots for if they become starved early in their life they will take quite a long time to recover. However, established ferns should not be over-potted as this can result in losses. Move them on into a slightly larger pot each time. Plants will need less frequent potting the larger they become and the larger the pots in which they are growing. The best time for potting ferns which are established is in March or April as they are starting to make new growth. The plants should always be moist at the time of potting and should be kept moist after they have been repotted.

6 Stem cuttings

This chapter and Chapters 7–12 are concerned with the second important method of propagation – that is, increasing plants vegetatively.
This is also known as asexual propagation and it differs from reproduction by seeds in that portions of plant growth are used to produce new
plants. Pieces of stem, roots, leaves, growth buds and portions of entire
plants can all be used in order to produce new plants.

Out of all the methods of vegetative propagation, the rooting of stem
cuttings is probably the most popular. Basically it involves inducing a
piece of stem or shoot to form roots of its own in suitable conditions, so
that eventually this shoot becomes a new plant, which is identical in
every respect to the parent plant. The reason why this is such a popular
method of plant raising is because so many plants can be increased
from stem cuttings: the list includes alpines, hardy perennials, half-
hardy perennials, shrubs, climbers, greenhouse plants, conifers, trees
and soft fruits of many kinds.

One of the great advantages of stem cuttings – and, in fact, all
methods of vegetative propagation – is that it can be guaranteed that
the new plants will be identical in every respect to the original or parent
plant. This is particularly important when we wish to propagate hybrid
plants and cultivars, as these will not come true to type when raised
from seeds.

More often than not stem cuttings root easily, provided they are
taken and inserted in the correct conditions. Research over many years
(it is still going on all over the world) has shown that even subjects that
were once considered difficult or impossible to root from stem cuttings
can now be rooted with no trouble at all if set in the optimum rooting
environment.

There are three groups of stem cuttings: softwood cuttings, semi-ripe cuttings and hardwood cuttings. It is a case of knowing which type is most suitable for the particular plant being propagated and this is indicated in the alphabetical tables on plant propagation which form the second part of this book. I will deal with each of these groups in detail.

Softwood cuttings

Softwood cuttings are prepared from current year's shoots before they become fully ripened, or hard and woody. As the term implies, the cuttings are soft or unripened. They are therefore collected and rooted in the early part of the year, mainly between April and June, although cuttings under glass can be rooted from January onwards if suitable shoots are available. Beyond June, though, plant growth starts to ripen and become woody and cuttings taken at this stage are termed semi-ripe cuttings.

Collection of cuttings Among the plants you may be propagating from softwood cuttings will be alpines, hardy and half-hardy perennials, shrubs, climbers and greenhouse plants. With many of these the material we are looking for is generally soft new side shoots. However with most hardy perennials or herbaceous plants (with the exception of monocotyledons, ferns and grasses) the best material consists of young shoots, about 5 cm (2 in.) long, produced from the crown of the plants. These are correctly known as basal cuttings and are preferably taken with a 'heel', so remove them as close as possible to the crown of the plant. The same applies to half-hardy perennials such as *Chrysanthemum* and *Dahlia*. Here, the dormant plants are started into growth in a heated greenhouse in January or February and the cuttings taken as soon as the shoots are of a sufficient length. Other plants that are started into growth early in the year to provide cuttings include greenhouse *Fuchsia*.

While on the subject of forcing plants into growth for cutting purposes, perhaps I should mention that there are various hardy shrubs that can be treated in this way in order to obtain really soft young growth. I must admit that this is mainly a commercial practice, but there is no reason why the amateur should not try it, for with certain subjects it results in better rooting. Young shrubs should be potted in the autumn and taken into a heated greenhouse early in the new year.

Provide only slight heat, say about 10°C (50°F) at the most, and this will result in soft lush shoots. Plants which root quite well from this type of treatment include deciduous azaleas (*Rhododendron*), Japanese maples (*Acer*), deciduous *Magnolia* as well as *Caryopteris* species and cultivars. As soon as cuttings have been taken the plants should be gradually hardened and then placed out of doors again. As I said, this is mainly practised by nurserymen, but it is an interesting method for anyone who likes to try something different. Of course, if you have only one plant of a particularly choice subject, you may be reluctant (very wisely) to lift it and force it into growth. There is no doubt that flowering would be affected that year.

Softwood cuttings are best collected early in the morning before the sun becomes too hot and the plants start to lose water and perhaps wilt. So try to collect the cuttings while they are 'fully turgid' – that is, full of water – and not when they are limp and the leaves are drooping. For best results it is essential to keep the cuttings turgid throughout the collection and preparation period; remove the complete shoot when collecting cuttings and pop into a clean polythene bag immediately, to prevent wilting.

I would not advise you to collect shoots which are infected with pests or diseases, as these will invariably be passed on to the new plants. In fact, some pests and diseases may even have an adverse effect on rooting.

It is better to prepare and insert the cuttings, working in a cool shady place under cover or indoors, immediately you have collected them, but if this is not possible they should keep in good condition for several hours if sealed in the polythene bag and kept cool. If the sun is allowed to play on them they will quickly wilt, even in the polythene bag, and even worse, they may be scorched.

Preparation of cuttings To prepare cuttings you will need a really sharp knife or even a razor blade, for the cuts must be clean and not ragged. Badly prepared cuttings are inclined to rot at the base instead of rooting, as ragged cuts do not heal, or are slower to heal over than clean cuts.

The shoots you have collected should be cut immediately below a node or leaf joint at the base, and these are correctly termed nodal cuttings. The tips of the shoots should be left intact. Cuttings of the majority of subjects will be in the region of 7.5 cm (3 in.) in length but with smaller subjects, like alpines for instance, the cuttings may be only

Figure 8 Preparation of softwood shrub cuttings. The shoot is cut immediately below a node or leaf joint at the base and the lower leaves are cut off

3.5–5 cm (1–1½ in.) long. The lower third or even half of the cuttings should be stripped of leaves by either cutting or pulling them off, taking care not to strip the bark. If leaves are allowed to remain on the lower part of the cuttings they will simply rot beneath the compost.

Cuttings of all plants will root more quickly and develop a stronger root system if they are treated with a hormone rooting powder prior to insertion. The lower 6 mm (¼ in.) should be dipped into the powder formulated for softwood cuttings, and the surplus needs to be removed by lightly tapping the base of the cutting on the edge of the container. (An excess of rooting powder on the base of the cutting may cause damage rather than stimulate rooting.) Contrary to popular belief there is no need to dip the bases of the cuttings in water before dipping them in the rooting powder, as there is sufficient moisture at the base to ensure that the powder adheres. (See Figure 9.)

Figure 9 Softwood cutting being dipped in a hormone rooting powder to ensure quicker rooting and a stronger root system

Inserting cuttings Cuttings of all subjects can be rooted in a cutting compost, which is easily mixed at home. It consists of equal parts by volume of moist sphagnum peat and coarse horticultural sand or grit. This will provide ideal rooting conditions: a well-drained and aerated compost which is able to hold sufficient moisture.

The cuttings should be rooted in containers of some kind, filled to within 12 mm (½ in.) of the top with cutting compost which should be moderately firmed. Suitable containers for large quantities of cuttings are standard seed trays about 5 cm (2 in.) in depth, but if you have only a few cuttings to insert then it would be more practical to use suitable sized plastic pots.

Cuttings are inserted in holes made with a wooden dibber (rather like a blunt pencil). Ensure that the base of each cutting touches the bottom of the dibber hole and that the lower leaves are just above compost level. The cuttings are firmed lightly with the dibber – or with the fingers if preferred – but it is particularly important to firm the cuttings at their base.

After insertion water the cuttings with a watering can fitted with a fine rose, allow them to drain and then put them in a suitable place for rooting.

Rooting conditions Softwood cuttings need heat and high humidity in order to root well. High humidity ensures that the cuttings do not wilt, which has the effect of considerably slowing down rooting. It is best, if possible, to provide heat at the base of the cuttings, where a

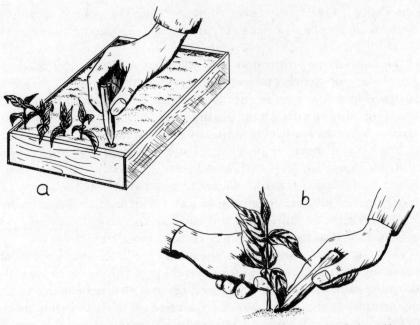

Figure 10 Inserting softwood shrub cuttings: (a) Use a dibber to make a hole and insert the cuttings almost up to the lower leaves. (b) Gently firm in each one with the dibber

temperature of 18°C (65°F) is desirable. If you are unable to provide bottom-heat then place the cuttings in a warm situation, with an air temperature again of 18°C.

There are many pieces of equipment available today which assist the rooting of cuttings. You can choose from a wide range of electrically heated propagating cases, for instance, which stand on a glasshouse bench and have electric heating cables in the base, controlled by a thermostat. To retain heat and ensure high humidity propagating cases are fitted with a transparent top or lid. These cases are quite cheap to run and it means that a high air temperature in the greenhouse is not necessary in order to root cuttings.

A more sophisticated piece of equipment is the mist-propagation unit, widely used by commercial nurserymen but also available in small versions for the amateur's greenhouse. A mist unit is installed on a glasshouse bench. Basically it consists of a number of spray heads which periodically spray the cuttings with a fine mist-like spray of water to prevent them from wilting. The spray heads run off the main water supply and are generally controlled automatically: as soon as the leaves of the cuttings start to dry they are sprayed with water. The water supply is generally operated by some kind of moisture-sensitive device which is placed among the cuttings and dries at the same rate as the cuttings.

Bottom-heat is provided by soil-warming cables laid in sand or shingle on the bench and controlled by a thermostat, so they are switched on and off automatically. You set the temperature required simply by turning a dial on the thermostat. (An electricity supply in the greenhouse is obviously vital to operate these cables!)

A mist unit makes it possible to root all kinds of plants – even 'difficult' ones like *Magnolia*, *Rhododendron* (especially the deciduous types) and *Cotinus*. I would not recommend that you root hairy or woolly leaved subjects under mist as water is inclined to collect on the leaves in excess quantities and this can lead to rotting. Such subjects are best rooted on the open bench of a heated greenhouse.

Of course a mist unit is not cheap to install, so another way of providing suitable conditions for the rooting of cuttings is to lay soil-warming cables in sand or shingle on the greenhouse bench. The air temperature in the greenhouse can then be kept fairly low (but above freezing) as you will have the heat where it is required – at the base of the cuttings. To ensure high humidity the cuttings can be rooted under a 'tent' of clear polythene sheeting supported on a light wooden or heavy galvanized-wire framework. The tent should be opened for an hour or so two or three times a week to avoid too much condensation, because over-moist conditions may result in the cuttings rotting. If the greenhouse is heated, this tent can, of course, be used even without soil-warming cables.

If you do not have a greenhouse, softwood cuttings may still be rooted if you have a warm windowsill indoors – preferably one which faces south. Insert the cuttings in pots and, in order to prevent them wilting, place a clear polythene bag over each pot with the edge tucked underneath. A polythene bag thus placed is inclined to collapse over the cuttings, so to prevent this happening two or three short split canes could be inserted in each pot to support the polythene. It is advisable to remove the polythene bag for an hour two or three times a week to avoid excess condensation. Alternatively a clear-plastic 'bell glass' could be used.

So often I hear of people despairing at the loss of their softwood cuttings because of rotting. This can be overcome in several ways. First, do ensure that the cuttings have sufficient heat and that the atmosphere is not over-moist – although humidity is necessary to prevent wilting.

Regular ventilation is vital to prevent a stale atmosphere. It is also essential to maintain hygienic conditions by using perfectly clean or sterilized containers plus a sterile compost of peat and sand. Always remove dead leaves (and cuttings) regularly to prevent diseases like botrytis or grey mould from building up which can lead to further rotting among your cuttings.

If you notice a good deal of rotting then it is a good idea to spray the cuttings with a fungicide such as benomyl. As an insurance policy, one can mix some fungicide with the rooting powder in which the cuttings are dipped – a suitable type is Captan. Many nurserymen even dip their cuttings in a solution of benomyl before inserting them as an insurance against rotting, and of course the amateur could also adopt this policy: the entire cutting is given a quick dip in the solution.

Aftercare of rooted cuttings How do you know when cuttings have rooted? Generally an indication is when the tips of the cuttings start to grow rapidly, although this is not an infallible guide. Another way of finding out is to inspect the bottom of the container – as soon as white roots start to show through then one knows that the cuttings have formed a substantial root system.

When well rooted the cuttings should be carefully removed from their containers and potted into 9 cm (3½ in.) pots, using a good potting compost such as John Innes No. 1 or an equivalent loamless type. Handle the cuttings carefully as very often the roots are soft and brittle and therefore easily damaged.

After potting, water them well and then return the young plants to the greenhouse until they have become established in their pots. If they are hardy or outdoor plants then you will need to harden them in a cold frame over several weeks to acclimatize them to outdoor conditions. After this hardening-off period the young plants will be ready for planting out in the garden, or maybe in a nursery bed to grow them to a larger size.

Semi-ripe cuttings

Semi-ripe cuttings are usually rooted in the period July to October, although some of the conifers are rooted as late as January or February. Semi-ripe cuttings are prepared from current year's shoots which are becoming ripe and firm (or woody) at the base, while the tops are still soft or unripened. Generally speaking this type of material is easier to root than softwood cuttings and a good many plants can be rooted quite easily without artificial heat. However, there are certain subjects which are more difficult and these need strong bottom-heat in order to root. The tables in the second half of the book give the relevant details.

A wide range of subjects can be propagated from semi-ripe cuttings, including shrubs, climbers, conifers, half-hardy perennials and greenhouse plants.

Collection of cuttings Wherever possible collect strong healthy side shoots from the plants you wish to propagate, removing the entire shoot. As with softwood cuttings ensure against wilting by placing the cuttings in a polythene bag, and collect them in the early morning when the shoots are full of water. In the summer you may find various pests and diseases on the plants – on no account prepare cuttings which are infected with either. It would be best to spray the plants some time before collecting cuttings to ensure clean, healthy material for propagation.

Preparation of cuttings Semi-ripe cuttings are generally nodal– that is, they are prepared by cutting cleanly immediately below a node or leaf joint at the base. Generally the tips of the cuttings are left intact, although with some subjects, like laurels (*Prunus*) and *Berberis*, I prefer to remove the soft tips. The lower third to half of the shoots should be stripped of leaves, either cutting or pulling them off. Only use the latter method if there is no risk of the bark tearing.

The length of cuttings will vary very much according to the subject. The average length for shrubs, conifers and so on is 10–15 cm (4–6 in.). Cuttings of heaths and heathers (*Erica* and *Calluna*) are very much shorter than this as they only make short side shoots, so these will be 3–5 cm (1½–2 in.) in length. The shoots of heaths and heathers should be pulled off the plants so that each cutting has a heel of older wood attached. The lower leaves are easily stripped off between finger and thumb, with the exception of *Calluna* which has very small scale-like leaves: these should be left intact. I generally pinch out the tips of heath and heather cuttings and it is certainly advisable to pinch out any flower buds as cuttings should not expend their energy in flowering.

There are various other subjects which require rather special preparation. For instance, some cuttings should be 'wounded' to encourage them to form roots, which involves removing a sliver of bark about 3

Figure 11 Preparing semi-ripe cuttings of *Pelargonium zonale*: (a) Cut the shoot immediately below a node or leaf joint and strip off the lower leaves. (b) Small quantities of cuttings can be inserted around the edge of a suitable size pot

cm (1½ in.) in length from the base of the cutting to expose the wood underneath. This treatment applies to plants which are more difficult to root, particularly holly (*Ilex*) and *Elaeagnus*.

I have already mentioned *Berberis* under straightforward nodal cuttings, but there are other ways of preparing cuttings of these shrubs. There is a special type of cutting known as a 'mallet' cutting, for instance, often used for *Berberis* which have thin shoots. Examples are *B.* × *stenophylla* 'Irwinii', *B. thunbergii* cultivars, and *B. wilsoniae*. Thin shoots may not root well and so they are taken with a portion of older wood attached at the base. Again use current year's side shoots but cut them off with about 6 mm (¼ in.) of the main woody stem attached. This will result in a cutting which looks rather like a mallet in shape, hence the term mallet cutting. I prefer to cut off the soft tip of each cutting. Also remove the lower leaves and the spines as this makes insertion much easier – although it is not a particularly pleasant or easy job removing needle-sharp thorns! (See Figure 12.)

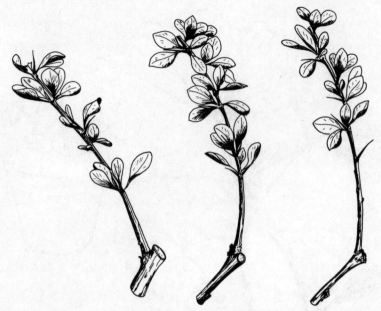

Figure 12 Mallet cuttings of a *Berberis thunbergii* cultivar. They consist of current year's shoots with a portion of older wood at the base

At the other extreme there are very strong-growing *Berberis* which produce shoots of about pencil thickness or more. Examples here are *B.* 'Chenaultii', *B. gagnepainii*, *B.* × *stenophylla*, and *B.* × *ottawensis* and cultivars. You can remove an entire one-year-old stem and, using a pair of sharp secateurs, cut it up into lengths of 15 cm (6 in.) to obtain several cuttings from one shoot. Discard the soft tip. The top of a cutting should be cut just above a node or leaf joint and the base just below a node. Again it is best to remove the lower leaves and spines from each cutting.

A word about conifer cuttings, which are straightforward nodal cuttings. Remember, though, that the base of each cutting should be well ripened and have at least 12 mm (½ in.) of brown wood. This means that the base is thoroughly ripened. You may find that the greater part of a conifer cutting is green and unripened, but as long as the base is ripe you can expect rooting to occur.

Having covered these various ways of preparing semi-ripe cuttings, do not forget that all will benefit from being dipped in a hormone rooting powder formulated for this type of material. Just dip the lower 6 mm (¼ in.) of each cutting in the powder and knock off the surplus.

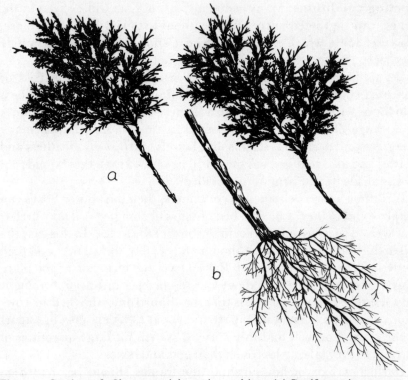

Figure 13 Cuttings of a *Chamaecyparis lawsoniana* cultivar: (a) Conifer cuttings are generally straightforward nodal cuttings, with a well-ripened base. (b) They generally root well without heat but rooting is very much quicker in heated conditions

Inserting cuttings Semi-ripe cuttings, like softwoods, are rooted in a cutting compost consisting of equal parts peat and coarse sand. Containers are also the same – pots can be used for small quantities of cuttings, and deep seed boxes for larger numbers. However, as we shall see under Rooting Conditions later in this section, some semi-ripe cuttings can be rooted in beds of cutting compost.

Whichever system of rooting you use, the cuttings should be inserted in holes made with a dibber and then moderately firmed in. Ensure the base of each cutting is in close contact with the bottom of the dibber hole and that it is inserted up to the lower leaves. When a batch has been dibbed in, water thoroughly with a rosed watering can to settle the cuttings.

Rooting conditions Many semi-ripe cuttings, as will be seen in the tables, can be rooted successfully without artificial heat. There are, however, some which benefit from heat and others which definitely need strong bottom-heat in order to form roots.

Let us first consider the rooting of semi-ripe cuttings in cold or unheated conditions. One of the most suitable structures would be a cold frame with a glass lid or top in a sheltered south-facing situation. A wide range of subjects can be rooted in this way, including many evergreen and deciduous shrubs, like laurels and *Berberis*; heathers and heaths; and also conifers, but these will be slow to root; they would root more quickly if some heat were provided.

If cuttings are to be rooted in containers, then the pots or boxes can simply stand in the frame on a bed of ashes or shingle. You may prefer, however, to root cuttings direct in the base of the frame. In this case the soil in the bottom of the frame should first of all be forked over and then moderately firmed with the feet. Rake it level and on top of the soil place a layer of cutting compost about 7.5–10 cm (3–4 in.) deep, firming it only lightly. The cuttings can then be dibbed into this bed in rows approximately 7.5 cm (3 in.) apart and about 5–7.5 cm (2–3 in.) apart in the rows. You will find this a useful system for large quantities of cuttings as they take up less room than pots and boxes.

During periods of hot sunshine the frames should be shaded to prevent the cuttings scorching. The frame lights should be kept closed to ensure a humid atmosphere within the frame. To maintain a higher temperature in the frame you could 'double glaze' the inside with polythene sheeting. Use clear polythene and either pin it to the under-side of the frame light or, better still, make up lightweight wooden frames to fit comfortably inside the frame, 'glaze' these with polythene and support them over the batch of cuttings. Such insulation will of course also help to protect the cuttings over the winter. Remember that many cuttings which are taken late in the year will spend the winter in their rooting environment.

Many semi-ripe cuttings can be rooted under low polythene tunnels. This applies to all those plants which can be rooted in a cold frame. A low polythene tunnel is just an alternative method and is a lot cheaper to construct than a set of cold frames.

You can buy low polythene tunnels complete in kit form. They consist of a strip of polythene (try to buy a tunnel in white polythene if possible) and galvanized wire hoops to support it. There may be a set of thinner wires to hold the polythene down.

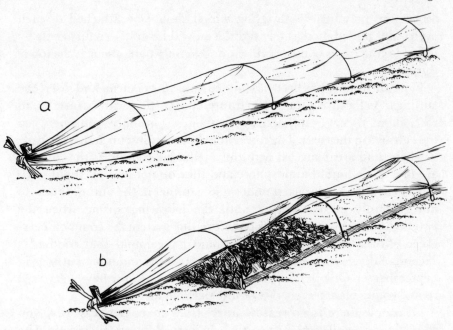

Figure 14 Rooting semi-ripe shrub cuttings under a low polythene tunnel: (a) The tunnel is 1 m (3 ft) in width and is constructed of wire hoops and, preferably, white polythene. (b) The cuttings can be rooted in a layer of cutting compost spread over the soil

Choose a well-drained sheltered site for your tunnel, for in a very exposed windy situation the polythene is liable to be blown off or torn. A warm sunny aspect is recommended as this will help to speed rooting.

The site should be prepared by first digging the soil to the depth of a spade or fork. Remove all weeds, especially perennial kinds. Then reduce the surface to a moderately fine tilth and tread firm. Rake to ensure a level surface. You will need to prepare a strip about 1 m (3 ft) in width, as most polythene tunnels are supplied in this width. Next a 7.5 cm (3 in.) deep layer of cutting compost is spread over the prepared strip and this can be topped, if desired, with a 2.5 cm (1 in.) deep layer of sharp sand to ensure good drainage around the cuttings.

The next stage is to insert the wire hoops, about 1 m (3 ft) apart down the length of the prepared strip, ensuring that the eyelets are at soil level. Next the polythene is placed over the hoops and the ends secured, either by burying them in the soil, or by tying them to short wooden stakes. The polythene should be pulled fairly tight. If you have thin wires for holding down the polythene, these should now be fitted; if not secure the polythene with string which should be attached to each eyelet and pulled moderately tight. You will be able to gain access by lifting up the edges of the polythene and loosely tying them at the top of the tunnel.

Once the tunnel has been assembled you can go ahead and insert the cuttings. As for direct rooting in frames, the cuttings can be inserted in rows about 7.5 cm (3 in.) apart with about 5–7.5 cm (2–3 in.) between the cuttings in the rows. The rows should preferably run across the bed.

I have had great success with this method of rooting as the cuttings are in a fairly humid atmosphere and they do not dry out rapidly. In fact, if cuttings are inserted under a low tunnel in the autumn they do not generally need watering until the following spring when the polythene cover is removed. I often use this system for common evergreen shrubs and also for *Berberis* and have found that rooting is exceptionally good. Generally I insert evergreens and other cuttings in September or October and by the following autumn have very well rooted young plants – very often quite sizeable specimens.

Of course there is no reason, apart from space perhaps, why you should not root all semi-ripe cuttings in heat. Rooting will certainly be quicker. But you may prefer to reserve heated space for those subjects which definitely need heat in order to root. I have already said that conifers will root without heat but it is a slow process so you may prefer to root these in heated conditions. Some of the more difficult subjects like holly (*Ilex*) and *Elaeagnus* will need fairly high temperatures in order to root well. All half-hardy plants will also root better with heated conditions.

Where heat is required, you can use a mist-propagation unit with bottom-heat; an electrically heated propagating case with bottom-heat; or a bench heated with soil-warming cables, as described in the section of this chapter on Softwood Cuttings. In the latter instance you could cover the cuttings with a polythene 'tent'. A suitable temperature for rooting cuttings is 18°C (65°F).

Aftercare of rooted cuttings Some cuttings will root within a matter of weeks, particularly half-hardy perennials and also heaths and heathers. Many cuttings taken early in the season will be well rooted by late summer or early autumn, but those inserted in late summer or in the autumn may not be well rooted until the following spring. Rooting time will also be influenced by temperatures, and those cuttings inserted in cold conditions will obviously be slower to root than those in heat. If I am rooting cuttings in the soil of a frame or under low tunnels, I prefer to leave them where they are until the following autumn, although they should have made a root system by the spring.

Once cuttings in cold frames have rooted the ventilation can be gradually increased by raising the frame lights. If under low polythene tunnels the edges of the polythene can be gradually raised to acclimatize the young plants to outside conditions. If cuttings are inserted under low tunnels in the autumn, the polythene can be removed in the following May when the weather should be getting warmer.

When you are satisfied that the cuttings are well rooted, they can be lifted. Cuttings raised in heat should be potted into small pots and then gradually hardened or acclimatized either to outdoor conditions in the case of hardy plants, or to the appropriate temperature in the case of greenhouse subjects. The hardening of hardy subjects can be done in cold frames.

If you have rooted cuttings direct in the base of a frame or under low tunnels they should make a substantial root system and could be planted in the garden or nursery bed once they have been well hardened. As already mentioned, I often prefer to leave the cuttings in the beds for maybe up to a year so that I can lift sizeable young plants. But you may prefer to clear your space for other uses.

I am sure that you will find the rooting of semi-ripe cuttings a fascinating subject and it is made all the more enjoyable by the fact that there are not so many problems as we have with the more delicate softwoods.

Hardwood cuttings

Quite a wide range of shrubs, trees, climbers and fruits can be propagated from hardwood or fully ripe cuttings. This is quite a straightforward method of propagation because preparation of cuttings is very simple and the majority can be easily rooted without artificial heat.

Collection of cuttings Hardwood cuttings are taken during November and December, after leaf-fall. It is best, in fact, to wait several weeks after leaf-fall as the cuttings will probably root better than if collected during or immediately after the leaves have fallen.

When collecting material for cuttings choose the current year's shoots which are well ripened, or hard and woody. You will need a pair of secateurs to remove the shoots.

Preparation of cuttings The shoots which have been collected are cut up into lengths of 15–25 cm (6–9 in.), depending on the subject, using a pair of sharp secateurs. Make the top cut just above a growth bud and the basal cut just below a bud. All the buds are left intact with the exception of gooseberries and red currants (*Ribes*). With these two fruits all but the top three or four buds should be cut out with a sharp knife to prevent the resultant bushes from producing growth from below ground level. Red currants and gooseberries are grown on a short leg or stem. With all other subjects we aim for growth from below ground level to ensure really bushy specimens.

Once the cuttings have been prepared the bases can be dipped in a hormone rooting powder which is formulated for hardwoods. Do not forget to tap off surplus powder on the edge of the container.

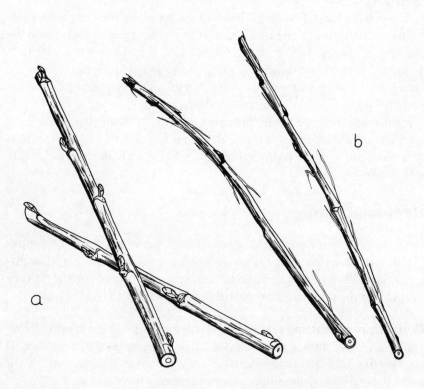

Figure 15 Hardwood cuttings of (a) black currant with all the buds left intact and (b) red currant with all but the top four buds removed

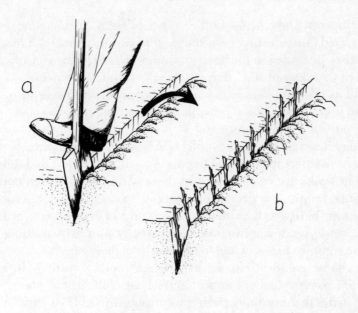

Figure 16 Inserting hardwood cuttings in the open ground: (a) First take out a
V-shaped trench. (b) Stand the cuttings in it as upright as possible. (c) Return the soil
and firm well in

Inserting cuttings Unlike other types of stem cuttings, hardwoods are inserted fairly deeply – to about half to two-thirds of their length. If they are to be rooted in the open ground or in cold frames it is the usual practice to take out a V-shaped trench, stand the cuttings in it, as upright as possible, return the soil around them and make firm. If the ground is not too hard it is possible and very much quicker just to push in the cuttings. (See Figure 16.)

If you intend rooting some subjects in heat, in a greenhouse, then you will need to insert the cuttings in pots of a suitable size. I find that 12.5 cm (5 in.) pots are convenient and these will hold about ten cuttings. Do not be tempted to overcrowd cuttings in pots. A suitable compost for rooting hardwoods in heat is a mixture of equal parts peat and coarse sand. Again you can insert the cuttings simply by pushing them into the compost to about half to two-thirds of their length.

This same compost can be used when rooting cuttings in a cold frame. Make up a bed of compost in the base of the frame, about 10 cm (4 in.) in depth and simply push the cuttings into it. If you do not want to make up a bed, then the cuttings could be inserted direct in the soil in the base of the frame, or you could use large pots if you have only small quantities of cuttings. Again use a peat/sand compost.

Rooting conditions Some hardwoods can be rooted in the open ground. Good examples are black and red currants, gooseberries, common hedging privet (*Ligustrum*), *Lonicera nitida*, willows (*Salix*) and poplars (*Populus*). It is only the easy rooters that can be set out in the open, but even with these it is highly desirable to choose a really sheltered site – one which is protected from cold searing winds which could dry cuttings and also scorch new growth in the spring. I would venture to say that outdoor rooting is probably more successful in the milder parts of the country.

It is also necessary to choose a site with very well-drained soil as heavy losses can occur if the ground lies very wet during the winter. Ideally a light sandy soil is best but of course not all gardens have this. One way to improve the soil and ensure better drainage is to dig in a good quantity of coarse sand or grit when preparing the site. Alternatively, coarse sand could be placed in the bottom of the V-trench when inserting the cuttings, to ensure free drainage at the base.

The warmer the aspect the quicker the ground warms up in the spring and the sooner rooting will commence. So try, if possible, to choose a sunny south-facing situation.

Hardwood cuttings in the open ground sometimes partially rise out of the soil when the ground is thawing after a hard frost. It is most important to go along the rows of cuttings re-firming them if this has happened, as they will not root if suspended in the soil with an air space below them.

Once the warmer weather starts in the spring it may be necessary to begin watering. This should be continued throughout the summer before the soil becomes too dry. Dryness will inhibit rooting so do keep the cuttings sufficiently moist at all times. Also carry out regular weed control to avoid the cuttings becoming smothered by weed growth, which will seriously affect the development of the top growth.

Many cuttings are best rooted in cold frames, to protect them from very wet conditions over the winter and from cold drying winds. The soil is generally better drained if the frame has been properly constructed. Typical examples of hardwoods which are best rooted in cold frames include *Forsythia*, shrubby dogwood (e.g. *Cornus alba* and cultivars), *Philadelphus*, *Weigela*, *Deutzia*, *Buddleia*, *Ribes*, *Spiraea* and *Viburnum*. Many of these have hollow or pithy stems which are more prone to rotting in wet conditions than very firm solid cuttings.

Again, the frames should be sited in a sheltered situation and in a warm, sunny, south-facing aspect. Once the cuttings have been inserted, either in a bed of cutting compost, or in the soil, or in pots, the frame lights should be placed over them and kept closed throughout the winter. By all means lift the lights periodically throughout the winter to make sure the cuttings are not drying and to remove any which may have died. About April or May, as soon as the days get warmer, the lights can be removed so that the cuttings are exposed to the open air throughout the summer.

Some of the trickier subjects need artificial heat in order to root well, which means placing them on a bench in a heated greenhouse, or on a bench equipped with soil-warming cables. A temperature at the base of the cuttings in the region of 18–21°C (65–70°F) would be ideal. I would not put these cuttings under polythene or under mist. If you have heat it will allow you to root such subjects as *Actinidia kolomikta*, *Celastrus orbiculatus*, *Ficus carica* (fig), *Metasequoia glyptostroboides*, *Parthenocissus* and *Wisteria*. More subjects are of course listed in the tables.

Aftercare of rooted cuttings Cuttings in the open ground, and in beds in a cold frame, should be left where they are until the following autumn by which time they will have formed a good root system and

81

can be lifted and planted elsewhere. They do not start to root until the spring, so do not be tempted to lift them at this time.

If the cuttings are in pots in a cold frame, you will have to lift the cuttings as soon as a reasonable root system has developed – once you can see white roots emerging through the holes in the bottom of the pots.

By the following autumn you will find that many have made really large specimens for planting out. This applies particularly to subjects like *Buddleia*, willows (*Salix*), and poplars.

When you place hardwoods in heat you will find that top growth is often very quickly produced – this of course is due to the temperature – but do not be in too much of a hurry to lift the cuttings for they will need several more weeks to produce a root system. Again you will know when an adequate root system has developed as the roots will show through the bottom of the pots. It will probably be early spring before the cuttings of most subjects are ready for lifting and potting into suitable size pots of John Innes or loamless potting compost. Once potted, return them to the greenhouse to become established and then transfer to a cold frame to harden them in readiness for planting in the garden or in nursery beds.

7 Other types of cuttings

Root cuttings

Increasing plants from root cuttings is, I find, one of the most enjoyable methods of vegetative propagation – possibly because it is so easy! All that is involved is chopping up roots into small sections and inserting them in compost; furthermore it is a most reliable way of increasing plants for up to 100 per cent success is often achieved, assuming suitable rooting conditions are provided. No sophisticated equipment is necessary as the cuttings can be rooted in cold frames, although some subjects do benefit from moderate heat such as is provided on a greenhouse bench.

A wide range of trees, shrubs, climbers, alpines and hardy perennials or herbaceous plants can be increased from root cuttings, as a glance through the tables will show. However, variegated forms will revert to green when propagated by this method.

Collecting cutting material Propagating from root cuttings means that the plants will have to be disturbed somewhat, and so the most suitable time of year for this method of increase is December and January, when the plants are dormant or in their resting stage. Try to choose a period when the ground is not too wet or frozen as this makes root collecting so much easier and the 'mother plants' will re-establish very much better.

The way in which the roots are collected will depend on the size of the plant. Small subjects, like alpines, hardy perennials and the smaller shrubs, can be completely lifted and then replanted as soon as a few roots have been removed. Large plants which cannot be lifted should have the soil scraped away until some suitable roots are exposed. Once

83

collected, the soil should be returned and firmed thoroughly with the heels – do ensure the plant is not left loose in the soil as it could be whipped around by the wind and be damaged or even die.

Most of the subjects which are propagated from root cuttings have thick fleshy roots, although there are some, like *Phlox paniculata* and *Primula denticulata*, which have thin wiry roots. When collecting from the thick-rooted subjects, such as *Rhus typhina* and *Papaver orientale*, choose roots of about pencil thickness. In all cases, however, choose only young roots for propagation as these will give a higher success rate.

On no account completely strip a plant of roots – generally about three, four or five from any one plant will provide sufficient cuttings for most people's needs. In any event leave the fibrous roots intact as then the plants will re-establish very much quicker– these are not needed for propagation in any case. Roots can be removed in their entirety without affecting the plant.

Always use a sharp knife or secateurs when removing roots to ensure clean cuts, which will heal very much better than ragged wounds made with a blunt knife. Ragged cuts are also more liable to become infected with diseases.

Try not to let the roots dry while collecting– it is best to collect into a clean polythene bag, especially if there are cold drying winds.

Preparing the cuttings Once the roots have been collected you can retire to the warmth of the greenhouse, potting shed or similar place under cover to prepare the cuttings in comfort. You will again need a very sharp knife or secateurs for this operation, to ensure that the cuttings do not rot due to infection by diseases.

I have already mentioned that some subjects have thick roots while others are thin rooted. The two types are prepared in slightly different ways. Let us first deal with the preparation of thick root cuttings.

These are cut into sections of about 5 cm (2 in.) in length and the top of the cutting should have a flat cut while the bottom should have a distinctly slanting cut. This is to ensure that the cuttings are inserted the right way up, for they are planted vertically. You may now be asking yourself: 'Which is the top and which is the bottom of a section of root?' The top of a root cutting is always that part of the root which was nearest to the stems of the plant, so bear this in mind when collecting and make sure you keep the roots the right way up. (The correct term for the top of a root cutting is the proximal end, while the bottom is known as the distal end.) Root cuttings will still 'take' if they are inserted upside down, but the shoots will start to grow downwards and will then turn upwards so that you have a kinked shoot which is very easily broken off when lifting the rooted cuttings. If a root cutting is inserted the right way up the shoot will grow straight and so is less liable to be snapped off.

Now let us consider the preparation of thin root cuttings. These are much easier to prepare as you do not have to worry about tops and bottoms – they are inserted horizontally – and all that has to be done is to cut the roots into 5 cm (2 in.) sections.

Once the cuttings have been prepared I generally like to dust them lightly with a fungicide, such as Captan, as I feel this does help to prevent rotting. My method is to put a very small amount of Captan dust in a polythene bag and then shake the root cuttings around in this so that they have a light coating of the fungicide. There is no need to use a hormone rooting powder as the majority of subjects root very easily without it.

Rooting conditions Most subjects will root well without the use of heat. A cold frame is a good place for most as the glass lights will prevent the cuttings from becoming excessively wet, which again can result in rotting. I know that some people root some of the tougher subjects in a well-drained spot out of doors (for instance *Rhus*, *Aralia* and *Eryngium*) but I still feel that protection from the worst of the winter

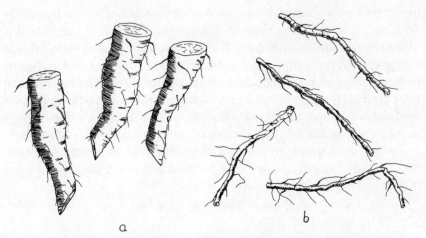

a   b

Figure 17 Preparation of root cuttings: (a) Thick root cuttings with a flat cut at the top and a slanting cut at the base. (b) Thin root cuttings. Both types are about 5 cm (2 in.) in length

wet is desirable. Even the lightest soils can become saturated in a very wet winter.

If you have a good number of cuttings to insert, it may be more convenient to root them direct in the base of the frame instead of in boxes or other containers. A 7.5 cm (3 in.) layer of cutting compost (equal parts sphagnum peat and coarse sand or grit) can be spread in the bottom of the frame, over the existing soil which should first of all be well broken up with a fork and moderately firmed. There is no need to firm the cutting compost; if it is left loose then it is easier to push in the cuttings.

For rooting small numbers of cuttings you may find it more convenient to use deep seed boxes of cutting compost in the frame. This is probably also a better method for thin root cuttings.

There are some subjects that root better in heat, such as *Romneya coulteri* and *Ceanothus*, as will be seen in the tables. These are rooted in deep seed boxes of cutting compost or individually in the modern unit containers – rather like seed boxes but divided up into small sections – which minimize root disturbance when lifting rooted cuttings. I generally stand the boxes on a greenhouse bench which is equipped with soil-warming cables, but they will be perfectly all right on a normal bench with a moderate air temperature. An electrically heated propagating case would be even better.

Insertion Once the cuttings have been prepared they should be inserted as soon as possible to prevent drying out. The thick cuttings are inserted vertically as already mentioned, making sure the flat cut is at the top and the slanting cut at the bottom. Simply push them into the compost so that the tops are slightly below the surface and then lightly cover the tops by brushing some compost over them. The cuttings can be spaced about 5 cm (2 in.) apart each way. If you are putting them in seed boxes you may find there is insufficient depth to insert the cuttings as suggested, and in this case they can be inserted at an angle of about 30°.

Thin root cuttings are laid on the surface of the compost, about 2.5 cm (1 in.) apart, and are then covered with a 1.2 cm (½ in.) layer of compost which should be lightly firmed.

Water the cuttings with a rosed watering can to settle the compost around them.

Aftercare Root cuttings should be kept moist but not excessively wet otherwise they may rot. If the cuttings have been inserted in cold

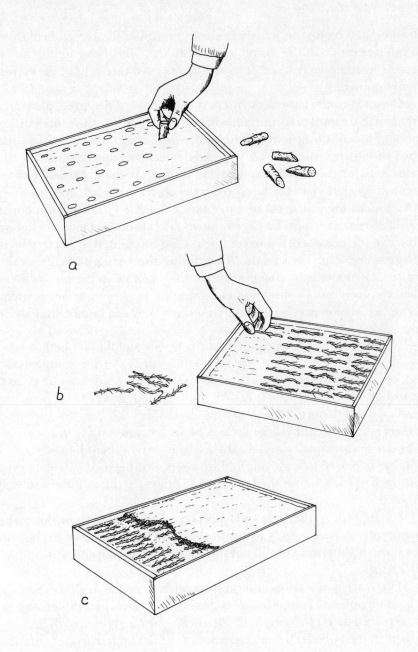

Figure 18 Inserting root cuttings: (a) Thick root cuttings are inserted vertically, with the flat cut at the top. (b) Thin root cuttings are laid on the surface of compost and (c) are covered with a 1.2 cm (½ in.) layer of compost

frames then retain the glass lights over the winter to protect them from winter wetness. In the spring, say about May, the frame lights can be removed completely as protection is not needed throughout the spring and summer.

Cuttings which have been inserted in the base of the frame generally remain there until early the following autumn. During the summer they should make substantial growth and will root into the soil below the bed of cutting compost, so you should have sizeable plants by the autumn. These should be carefully lifted with a garden fork – remember that you may have to dig deeply as many are deep-rooters. Once lifted the young plants can either be planted in their permanent situations or, if you prefer to grow them to a larger size before including them in the ornamental garden, they could be planted in nursery beds to grow for a further season. When lifting the young plants from the frame, try not to leave any portions of roots behind in the soil, as these will result in more young plants coming up between any subsequent crops which you may have in the frame and they can be as troublesome as weeds.

If you root the cuttings in seed boxes, they should be lifted as soon as they have rooted: there are no plant foods in cutting compost and so the young plants would quickly starve. I should say that root cuttings, especially those raised in a heated greenhouse, produce top growth before they start rooting. This is due to the food reserve in the cuttings. So do not be in too much of a hurry to lift the cuttings when you see shoots and leaves being produced. I find it is better to wait until white roots can be seen just emerging through the bottom of the boxes – this indicates the cuttings are well rooted.

It is best to tap the sides and ends of the box on a bench to loosen the compost, and then gently throw the batch of cuttings out of the box on to a bench. Alternatively lift out the young plants carefully with a small hand fork.

There are now various alternatives open to you as regards growing the young plants. I would suggest they are potted into suitable size pots and placed in a cold frame to harden before being planted in the garden or into nursery beds. If you do not have facilities for hardening, then the young plants could go straight out into the garden or nursery beds, but this is not ideal as they may receive an initial check to growth. Use a good potting compost, such as John Innes potting compost No. 1 or an equivalent loamless type.

Leaf cuttings

Certain plants can be propagated from leaves used as cuttings. When inserted the leaves form roots and eventually young plants. It is mainly greenhouse or pot plants that are increased by this method although there are also a few hardy plants that respond to this type of propagation, as can be seen in the tables.

Leaf cuttings are generally taken in the spring and summer and they usually need heat in order to root, so use a propagating case or a bench in a heated greenhouse.

Leaf cuttings can be prepared in various ways, according to the plant, so let us look at the various methods in detail.

Preparation and insertion Sometimes whole leaves are used, complete with the leaf stalk. This applies to *Peperomia, Saintpaulia*, and the rock plants *Haberlea, Lewisia* and *Ramonda*.

Use a very sharp knife to remove the leaves from the parent plants and then dip the ends of the leaf stalks in a hormone rooting powder which is formulated for softwood cuttings. Do not forget to knock off the surplus powder. The cuttings are then inserted in a cutting compost, the entire leaf stalk being below the surface of the compost. Use a dibber to make a hole for each cutting and lightly firm each one with the fingers. (See Figure 19.)

Small quantities of cuttings can be inserted around the edge of a suitable-sized pot, but for larger numbers use a standard seed box.

There are one or two species of greenhouse *Begonia* which can be propagated from leaves – typical examples are *B. masoniana* and *B. rex*. Here we take an entire leaf again – a fairly large one – but this time remove the leaf stalk. Now turn the leaf upside down and with a sharp knife slice through the main veins in a number of different places. Make sure you cut right through the veins.

Now the leaf is laid on the surface of cutting compost in a seed box, top side facing upwards. It is important to ensure that the cut veins are in close contact with the compost; this can be achieved by weighting the leaf with a few small stones, or some small wire pegs could be pushed through the leaf to hold it in close contact with the compost. Roots will be produced where the veins have been cut, and after this new plants will develop on the upper side of the leaf.

With subjects like *Sinningia* (gloxinia) and *Streptocarpus* entire leaves are used but, because of their length, we generally cut them in half and

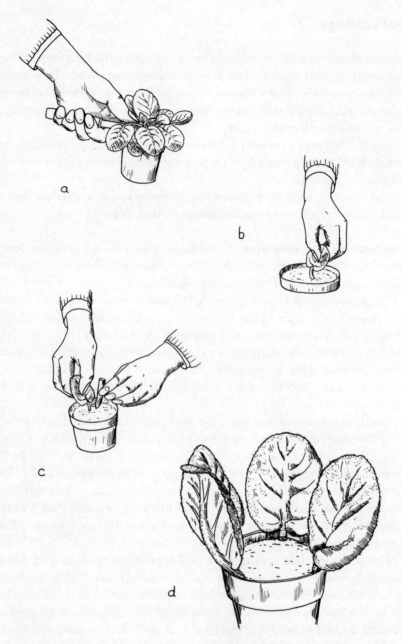

Figure 19 Leaf cuttings of *Saintpaulia*: (a) Leaves are cut off complete with petiole or leaf stalk. (b) They are dipped in a hormone rooting powder. (c) The cuttings are inserted up to the base of the leaves. (d) Small quantities of cuttings can be inserted around the edge of a pot

use the lower half as the cutting. These are treated with hormone rooting powder and then inserted upright, either in pots or seed boxes.

The leaves of *Sansevieria* are cut into 5 cm (2 in.) long sections. This plant has long, stiff, sword-like leaves so remove an entire leaf and cut it into a number of sections. This method is used for *S. trifasciata* and other species, but do not propagate the yellow-edged *S. trifasciata* 'Laurentii' by this method as the result will be all-green plants – the yellow edge to the leaves just does not remain in plants propagated from leaf cuttings. Use division as a method of increase instead (see Chapter 10).

When the cuttings have been prepared, dip the bases in a hormone rooting powder and insert them vertically (making sure they are the right way up) to half their length in pots or boxes of cutting compost. To ensure that the cuttings are inserted the right way up, cut the leaf on a bench and lay out the cuttings in the right position, because the more they are handled before insertion, the more chance there is of confusion.

There are a number of succulent plants that can be increased from leaves and these include *Aloe, Crassula, Echeveria, Mesembryanthemum* and *Sedum*. Use whole leaves as cuttings and insert them in an upright position, shallowly, in pots or boxes of cutting compost.

There are other succulents which produce tiny plantlets on the edges of their leaves – these naturally drop to the soil and root. *Bryophyllum* of various species have this method of vegetative increase. Instead of waiting for the young plants to fall and root, it is possible to remove an entire leaf complete with plantlets and to lay this on the surface of the cutting compost: the plantlets will quickly root and start to grow.

Rooting the cuttings It is best to provide a temperature of 18°–21°C (65°–70°F) to ensure that leaf cuttings of all types root successfully. High humidity is also desirable. The best place is therefore an electrically heated propagating case, or the cuttings could be rooted in a mist-propagation unit. Another method is to stand the cuttings on a greenhouse bench and cover them with a polythene 'tent', or pots of cuttings could be placed on a windowsill in a warm room indoors and enclosed in a clear polythene bag. If cuttings are under polythene, remove this for an hour or so two or three times a week to avoid excess condensation otherwise the cuttings may rot.

The young plants Once leaf cuttings have rooted and the young plants have formed, they should be carefully lifted and potted into 7.5 cm (3 in.) pots, using either John Innes potting compost No. 1 or the

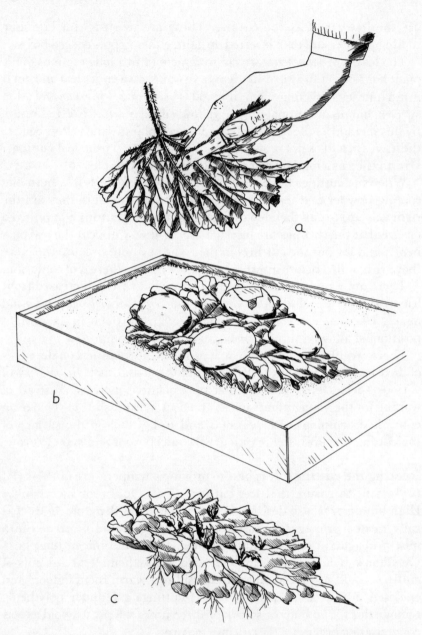

Figure 20 Leaf cutting of *Begonia rex*: (a) Take a fairly large leaf and cut through the main veins in a number of places on the underside. (b) Lay the leaf on the surface of the compost, top side facing upwards, and weigh it down with a few small stones. (c) Where the veins have been cut roots will be produced and after this new plants will develop on the upper side of the leaf

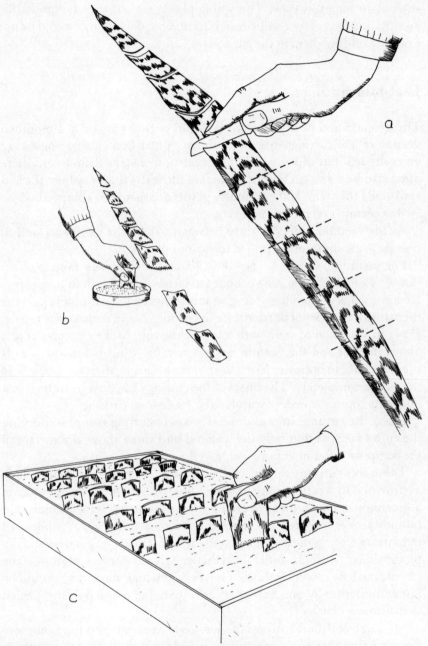

Figure 21 Leaf cuttings of *Sansevieria trifasciata*: (a) An entire leaf is cut into a number of sections. (b) The bases are dipped in a hormone rooting powder. (c) They are pushed into the compost, to about half their length, making sure they are the right way up

equivalent loamless type. The young plants should then be gradually acclimatized to cooler conditions. If hardy, harden them in a cold frame prior to planting them in the garden.

Leaf-bud cuttings

The propagation of plants from leaf-bud cuttings is really a modified version of leaf cuttings described above. A leaf-bud cutting consists of an entire leaf but this time we take a portion of stem with it, and there must also be a growth bud in the axil of the leaf– that is, where the leaf stalk joins the stem. Various hardy plants are raised by this method, as well as certain greenhouse subjects.

As the method differs slightly between subjects it will be as well to discuss the technique as applied to various plants.

For instance, some plants, like *Clematis*, ivy (*Hedera*) and passion flower (*Passiflora*) should be propagated from soft growth in the spring. Remove a few of the long young shoots from the parent plants and cut them into a number of portions. Each cutting should consist of a 1–2 cm (½–1 in.) portion of stem with a leaf at the top. Make the top cut just above the leaf and the bottom cut 1–2 cm (½–1 in.) below the leaf. If the leaves are in opposite pairs, as in some subjects, then remove one of the leaves completely. The bases of the cuttings can now be dipped in a hormone rooting powder formulated for softwood cuttings.

Insert the cuttings in pots or seed boxes of cutting compost, burying the entire stem so that only the leaf and bud show above the surface of the compost. After insertion, water well.

These soft cuttings should be rooted in warm moist conditions such as provided by a propagation case or a mist unit, or they can be stood on a greenhouse bench and covered with a polythene 'tent' to maintain humidity. Yet another alternative is to root them on a windowsill indoors, enclosing the pot in a polythene bag. Do not forget periodically to ventilate cuttings under polythene. Given warm conditions the cuttings will be rooted within a few weeks at which stage they should be potted individually and hardened prior to being planted in the garden or in nursery beds.

Although leaf-bud cuttings of *Camellia* are prepared in the same way use semi-ripe current-year's stems; a good time of the year to root them is in August. Use a stronger hormone rooting powder this time – one formulated for semi-ripe cuttings. I should not hide the fact that

Camellia may not be the easiest subject for the amateur gardener to propagate but it is well worth having a try if you can provide a high bottom-heat temperature. So a propagating case with soil-warming cables in the base is desirable; alternatively a mist-propagation unit with bottom-heat. Nurserymen often propagate *Camellia* from leaf-bud cuttings as they are especially economical of propagation material – the amateur gardener may also wish to economize, especially if he or she has only a small plant which it is desired to increase.

Camellia cuttings may take about eight weeks to form a good root system, after which they should be potted into individual small pots, using an acid compost, and overwintered in cold frames or in a cold or slightly heated greenhouse. They will in fact make faster growth if they are given a complete growing season in a cold or cool greenhouse. Then in the late summer they can be hardened ready for planting in the garden.

Mahonia can also be propagated from leaf-bud cuttings in October or April; autumn is preferred. Again use semi-mature stems and make sure they are green and not brown; the latter denotes fully ripened or hardened wood which may not root. As they are rather hefty, cut the portion of stem to about 3.5–5 cm (1½–2 in.) in length and reduce the length of the leaf to about three pairs of leaflets. *Mahonia* cuttings need bottom-heat for rooting and I normally cover the cuttings with polythene to maintain a really humid atmosphere. Rooting is fairly slow but as soon as a good root system has been formed pot the cuttings and place in a cold frame to thoroughly harden before subjecting them to outdoor conditions.

August is a good time to prepare leaf-bud cuttings of blackberries and loganberries (*Rubus*) and these are easily rooted in a cold frame in seed boxes of cutting compost. In the following spring the rooted cuttings can be lifted and either planted out or potted to grow on.

The greenhouse plants that are increased from leaf-bud cuttings are the rubber plant (*Ficus elastica*) and *Dracaena*. Propagate from young growth during spring or summer. Rubber plants have large leaves and therefore it is the usual practice to roll the leaves longitudinally into a cylinder and secure them with an elastic band. This way they take up far less room. Also each cutting will need supporting with a thin cane so that it remains in an upright position. I suggest that you insert one cutting per 7.5 cm (3 in.) pot, using cutting compost. *Dracaena* leaf-bud cuttings will also need supporting with a cane and again they can be inserted in individual pots. Both subjects will need warmth and humidity

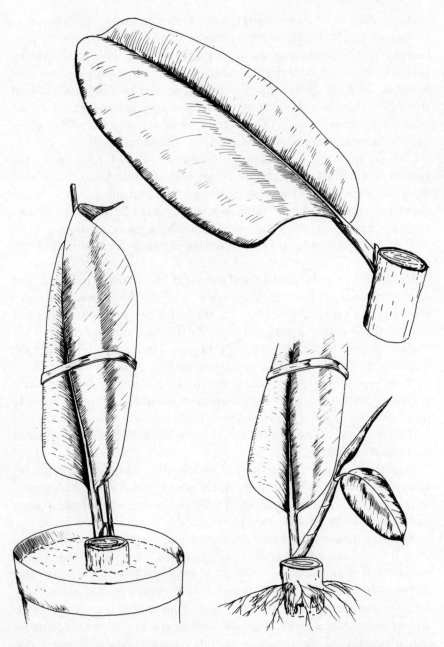

Figure 22 Leaf-bud cuttings of *Ficus elastica*, the rubber plant. Each cutting consists of a 2.5 cm (1 in.) portion of stem and a complete leaf with a growth bud in the axil. Insert one cutting per pot, roll the leaf into a cylinder and secure with an elastic band, and provide a thin cane for support. Warmth and humidity are needed for rooting

1 Cones of *Cedrus,
Abies, Sequoiadendron*
and *Pinus*. The scales
of most of these have
opened to release the
seeds, the exception
being the *Cedrus* (top
left)

2 Seed heads of
Clematis (left), *Corylus*
(centre) and *Laburnum*
(right)

3 Seed heads of *Cercis*
(left), *Piptanthus*
(centre) and *Hydrangea
paniculata* (right)

4 *Cotoneaster* berries (left), *Ceanothus* seed head (centre) and *Pyracantha* berries (right)

5 *Rosa moyesii* heps (left), *Rosa rubrifolia* heps (right) and *Leycesteria* seed head (centre)

6 Seed heads of *Papaver* (top), *Lathyrus* (centre) and *Crocosmia* (bottom)

7 Seed heads of *Kniphofia* (left) and *Acanthus* (right)

8 Seed heads of *Eryngium* (left), *Achillea* (centre) and *Lupinus* (right)

9 Seed heads of *Phlomis* (left), *Limonium* (centre) and *Malva* (right)

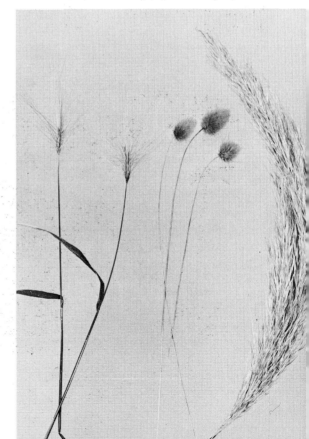

10 A selection of grass seed heads

11 Seed heads of *Galtonia*
(left), *Allium* (centre) and
Agapanthus (right)

12 Seed heads of *Incarvillea*
(left), *Callistemon* (centre)
and *Iris* (right)

13 Collecting seeds of *Sorbus*. Note that the entire bunch of berries is being picked off the tree

14 Collecting seed pods of a hybrid *Lilium*

15 The use of a seed sieve to
separate seeds from chaff

16 A mixture of large seeds and chaff (crushed seed capsules)

17 Separating large seeds and chaff by hand

in order to root and treatment with a hormone rooting powder formulated for softwood cuttings. Once the small pots are full of roots the plants can be potted into a size larger pot using a good potting compost.

Eye-cuttings

Fruiting and ornamental vines can be propagated from eye-cuttings during December and January. Basically an eye-cutting is a short piece of stem with a dormant growth-bud or 'eye', and these cuttings are rooted in heat.

The fruiting grape vine, *Vitis vinifera*, and its cultivars like 'Black Hamburg', 'Buckland Sweetwater' and 'Purpurea', respond well to this method of propagation. I also root other, ornamental, vines in this way, such as 'Brant', *Vitis coignetiae*, *V. davidii* and *V. pulchra*, which are often grown for their good autumn leaf colour.

If sufficient heat is provided a high percentage 'take' can be expected. It is best to provide bottom-heat for the cuttings as for example in a propagating case or on a greenhouse bench with soil-warming cables. A temperature in the region of 21°C (70°F) is desirable. I find it is not necessary to provide mist or to cover the cuttings with polythene but if a propagating case is used then of course the lid should be closed to conserve heat.

Preparation and insertion The material used for eye-cuttings should be well-ripened hard stems which were produced in the previous growing season – that is, one-year-old wood. Use a pair of really sharp secateurs for the preparation of these cuttings as the stems are often quite tough, and cut them into sections about 2.5–3.5 cm (1–1½ in.) in length. Make the top cut just above a bud or node, and the bottom one between buds. Leave one bud at the top of the cutting and carefully cut out the opposite bud. (See Figure 23.)

Another way to prepare eye-cuttings is to cut the stems into 3.5 cm (1½ in.) lengths, each with a growth-bud or eye in the centre. A sliver of wood can be removed on the opposite side to the bud to ensure quicker rooting. The choice is up to you – good results can be obtained with either approach.

I usually prefer to treat the cuttings with a hormone rooting powder recommended for hardwood cuttings: there is no doubt this speeds rooting.

97

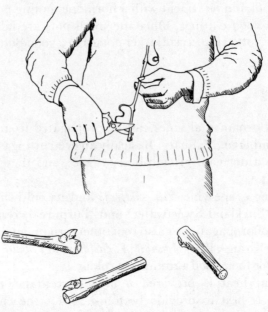

Figure 23 Eye- cuttings of grape vine (*Vitis*). They should be prepared from well-ripened shoots, which are cut into sections about 2.5–3.5 cm (1–1 ½ in.) long with an eye or bud at the top.

Eye-cuttings are rooted in cutting compost and can either be inserted in standard seed trays or in individual small plastic pots or other containers like compressed-peat pellets. Inserting a single cutting into a small container will minimize root disturbance when it comes to potting the rooted cuttings. If compressed-peat pellets are used there is no need for a cutting compost. Small compressed-peat pots also make good containers as again the rooted cutting can be transferred from its container without disturbance.

The method of insertion will depend on the way in which the cuttings have been prepared. If there is a bud or eye at the top of the cutting, then it can be inserted vertically, pushing it into the compost so that the bud is just showing above the surface. If your cuttings have a bud in the centre, then they should be inserted horizontally. Simply press them into the compost so that the bud faces upwards and is above the level of the compost, firming just sufficiently to ensure the cuttings remain stable.

Water the cuttings to settle the compost around them and then place in the propagating case or on the greenhouse bench.

Figure 24 Inserting eye-cuttings. Only the top bud needs to be exposed above the compost. Simply push in the cuttings to the correct depth

Aftercare You will find that eye-cuttings quickly produce top growth from the bud or eye – this is due to the warm conditions provided – but do not make the mistake of thinking that the cuttings have formed an adequate root system at this stage. It will be early spring before they are ready for potting.

If you rooted the cuttings in seed boxes then they will have to be carefully lifted and potted into suitable-sized pots, using John Innes potting compost No. 1 or an equivalent loamless type. If they were rooted in individual small plastic pots, tap out each one and pot into a size larger container, using the same compost. Rooted cuttings in compressed-peat pellets or compressed-peat pots are not removed from these but are potted, complete with peat-container, into a suitable size of plastic pot. The peat-containers gradually disintegrate and will not affect root growth in any way.

It will now be necessary gradually to harden the rooted cuttings. They can be returned to a greenhouse until established in their pots, and then transferred to a cold frame for gradual acclimatization to outdoor conditions. Do not forget to provide a cane to support these subjects, for remember they are climbers and produce long stems. Once hardened, either plant in the garden or in nursery beds to make larger specimens. You will find that most vines will have made fairly large plants within about eighteen months from propagation, as they have a vigorous habit of growth.

Pipings

Pinks (*Dianthus*) are propagated from a type of cutting known as a 'piping'. All the perpetual-flowering border pinks, such as the *Allwoodii* varieties, can be increased by this method, as can the many kinds of alpine or rock-garden pinks.

The best time to propagate pinks from pipings is in July and August, once the plants have made sufficient growth. Choose strong shoots which have not become unduly long. To prepare pipings, simply pull out the top of a young shoot – it should come out cleanly, snapping off just above a node or leaf joint. There is no need to use a knife. The length of piping will vary according to the variety or species, and those of some alpine kinds will be quite short. However, try to ensure the pipings have about three or four pairs of well-developed leaves after the lower pair has been pulled off. Generally speaking the older varieties of pinks root better than the more modern varieties, so if you experience difficulty with the latter then propagate them from conventional cuttings, using a sharp knife and making the bottom cut just below a leaf joint or node.

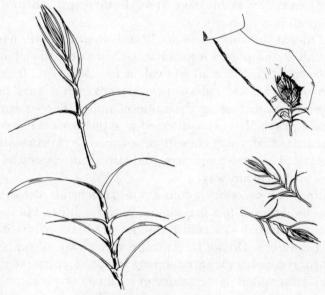

Figure 25 Pipings of pinks (*Dianthus*). These are a type of cutting. To prepare them, simply pull out the top of a young shoot so that it snaps off cleanly at a node or leaf joint. Remove the lower pair of leaves

The pipings can be inserted in 9 cm (3½ in.) pots; nine to twelve per pot will be ample to prevent overcrowding. There are various types of rooting media one could use: coarse sand, for example, or a mixture of three parts sand and one part Vermiculite. Insert the pipings up to the base of their lower leaves, taking care not to bury the bases of these leaves. After insertion, water the pipings and then place the pots in a cold frame.

The frame lights should remain closed for several days to maintain a humid atmosphere after which time a little ventilation should be given.

Within about three to four weeks the pipings (or cuttings if this method has been used) will have formed roots. At this stage start to provide more ventilation, increasing it over the next ten days before removing the lights completely. By this time the pipings or cuttings will be sufficiently well rooted to be either potted into 9 cm (3½ in.) pots of John Innes potting compost No. 1, or planted in the garden or in nursery beds.

Finally it is worth mentioning that it is a good idea to produce a steady succession of young pinks to replace older plants on a regular basis: you will obtain a far better flower display in this way. Pinks start to decline in vigour and flower less freely the older they are, and they also become rather woody and straggly. The plants could well be replaced about every three to four years, before they reach this stage.

Rooting cuttings in water

There may be many readers who do not have greenhouses, frames, propagating cases and so on in which to root cuttings. Nevertheless, cuttings of many plants can be rooted in water on a warm windowsill indoors. This is an especially suitable method of taking cuttings for those people living in flats; and it is fascinating for children to see developing roots.

There are a great many subjects that would respond to this method and readers could well try out a wide range of plants. I have had success with the following greenhouse plants: *Codiaeum*, *Coleus*, *Fuchsia*, *Impatiens*, *Nerium oleander*, *Tradescantia* and *Zebrina pendula*. Some of the easier hardy plants also respond to this method, particularly the willows (*Salix*).

Propagation is best carried out during warmer weather, in the spring and summer, using young shoots. The cuttings are prepared in the normal way: that is, they can be made approximately 7.5–10 cm (3–4 in.) long, using the tips of shoots, and cutting the base just below a node or leaf joint. Remove the lower leaves as they will rot under water and turn the water stagnant.

Now stand the cuttings in a small or shallow jar and add just sufficient water to cover the stripped stems. Place a piece of charcoal in the bottom of the jar to prevent the water becoming sour.

Once roots start to form – say when they are about 2.5 cm (1 in.) in length – the cuttings should be potted individually into 7.5 cm (3 in.) pots of potting compost. Do not let the roots grow too long before potting as they will become entangled and difficult to separate. They will also be rather brittle and liable to damage, and thus slow to establish in their pots.

Irishman's cuttings

These are partially rooted shoots which are carefully pulled away from plants and planted elsewhere to provide new plants. It is only possible to use this type of material with plants which produce shoots from their base or from the root area, such as Michaelmas daisies (*Aster*), *Chrysanthemum* and many alpines or rock plants. It is a useful method if you want only one or two new plants, or a friend or neighbour requires a piece of some particular plant. Carefully remove a few rooted shoots any time in the growing season, and allow them to become established in pots in a cold frame prior to planting in the garden.

This is a particularly good way to propagate pansies (*Viola*); late summer is a suitable period. The best approach is to cut back the topgrowth of the plants and to place some potting compost or fine garden soil around and between the stems. This will encourage roots to develop at the base of the stems. The rooted stems can then be cut off and should be planted in a cold frame; or they can be potted and stood in the frame. In the following spring put the young plants in the garden. Quite a few new plants can be obtained from one parent plant and pansies need replacing regularly with new plants as they are short-lived perennials and soon deteriorate in quality.

8 Grafting

Grafting is one of the most skilled aspects of propagation for it involves very careful and accurate use of the knife. Many amateurs are put off this method of producing new plants for this reason, and prefer to leave it to nurserymen who are adept at knifework. But I would quickly add that the keen amateur should try, for great satisfaction can be achieved from this method of propagation. There is no doubt that it will take practice, and the first attempt may result in complete failure. But, as with any other skill, it can be mastered with practice.

Grafting is a matter of forming a permanent union between portions of two separate plants so that they develop into a complete new plant. The root system, known as the rootstock, is provided by one plant and a piece of shoot or stem of the plant you wish to propagate is grafted onto this rootstock to form the top growth of the plant – the stems, foliage and flowers. This piece of stem is correctly called the scion.

You may now be asking yourself – what is the purpose of grafting? There are several reasons why this method of propagation is used. For instance, some plants are difficult to increase from cuttings – indeed there are many trees which simply will not root from cuttings. You may say, then why not layer them? But there are certain subjects, particularly fruits like the apple (*Malus*), cherries, peaches, plums and so on (*Prunus*), and the pear (*Pyrus*) which do not grow well on their own roots. They need special rootstocks to ensure adequate growth and fruiting. The fruit trees which we buy from nurserymen are on such rootstocks. Very often the rootstocks for fruit trees are specially chosen to control the growth and eventual size of the tree. With apples, for instance, there are some rootstocks which produce dwarf trees (like bush trees, cordons and espaliers), others which result in medium-sized trees and yet others which produce very vigorous growth, resulting in really large trees.

Very often one can obtain plants very much quicker by grafting than with other methods of propagation. For instance, with many trees it is possible to get new plants up to a height of 1.5–2 m (5–6 ft) in less than a year from grafting.

Generally speaking one usually grafts only cultivars or hybrids because these do not breed true from seeds. Otherwise seeds would provide a suitable method of propagation for certain trees. In fact, I suggest that if you wish to propagate species of trees (which breed true to type) then you should raise them from seeds rather than by grafting.

Before going into detail about the various methods of grafting it must be stressed how absolutely essential it is to use a razor-sharp knife to produce good clean cuts. You will also find the operation very much easier if the knife is sharp. If you intend doing very much grafting then it would be a good idea to invest in a proper grafting knife – a fairly heavy knife with a strong straight blade – for the type of grafting known as 'whip-and-tongue'. For the other forms of grafting, which will be discussed later in this chapter, you can get away with a budding knife, which is much lighter in weight. A budding knife, apart from being useful for some types of grafting (and of course budding – see Chapter 9) can also be used for the preparation of cuttings and other aspects of propagation.

I would suggest knives with good-quality steel blades which can be sharpened to a keen edge. Good steel will remain sharp for much longer than a knife with a poor-quality blade. I certainly would not recommend a knife with a stainless-steel blade as it is very difficult to achieve a really sharp cutting edge. To summarize, when buying knives choose the best that you can afford – the more you pay the better the quality of the blade.

There are many kinds of graft; in fact there is even a book devoted entirely to this subject. But we can reduce the number to four or five – even the commercial nurseryman needs no more than this. So the types of graft that I will discuss are whip-and-tongue, which is carried out in the open ground; and 'saddle', 'spliced side', 'veneer' and 'spliced side veneer', which are done in heated conditions.

Whip-and-tongue grafting

This is a very common form of grafting and is used to produce many kinds of fruiting and ornamental trees. It is performed in the open ground during the period February/March. I would advise you to choose a warm day, if possible, when carrying out whip-and-tongue grafting: it is very difficult to control the knife with cold hands – and of course there is a greater risk of cutting yourself.

Rootstocks – how to obtain and grow them In grafting it is usually necessary to ensure that the plant to be propagated is closely related to the rootstock on which it is to be grafted. More often than not the rootstock used is the common counterpart of the cultivar or hybrid. For instance ornamental crab apples can be grafted on to the common or wild crab apple, and cultivars of mountain ash worked on the common mountain ash, *Sorbus aucuparia*.

The following table lists trees which are generally grafted and the rootstock which should be used for each.

Plant to be propagated	Suitable rootstock
Acer platanoides cultivars (Norway maples)	*Acer platanoides* (Norway maple)
Acer pseudoplatanus cultivars (sycamores)	*Acer pseudoplatanus* (sycamore)
Crataegus cultivars (thorns)	*Crataegus monogyna* (hawthorn)
Fagus sylvatica cultivars (beeches)	*Fagus sylvatica* (common beech)
Laburnum cultivars	*Laburnum anagyroides* (common laburnum)
Malus, fruiting cultivars of apple	Malling Merton and Malling apple rootstocks
Malus, ornamental crab apples	As above, or *Malus sylvestris* (common crab apple)
Prunus, almonds	Plum rootstock Brompton
Prunus, apricots	Plum rootstock Brompton
Prunus, cherries, both fruiting and ornamental	*Prunus avium* (wild cherry)
Prunus, nectarines and peaches	Plum rootstock Brompton
Prunus, plums	Plum rootstock Myrobalan B
Pyrus, pears, fruiting and ornamental	Malling Quince A (selection of common quince, *Cydonia oblonga*)
Robinia pseudoacacia cultivars (false acacia)	*Robinia pseudoacacia* (false acacia)
Sorbus aria cultivars (whitebeams)	*Sorbus aria* (common whitebeam)
Sorbus aucuparia cultivars (mountain ash)	*Sorbus aucuparia* (common mountain ash)

The next question is how to obtain suitable rootstocks. In certain instances this could prove somewhat difficult for the amateur gardener, particularly with the special Malling Merton and Malling apple rootstocks. These stocks were originally produced at research stations, after which they are named, and there is a range of numbered rootstocks for definite purposes – for example, to produce dwarf trees, medium-size trees and very large vigorous trees. Generally these are only available in quantity to the nursery trade. The same applies to the plum rootstocks Brompton and Myrobalan B, and to Malling Quince A which is used as a rootstock for pears. You may, of course, know of a nurseryman who could let you have a few of these particular rootstocks, but on the other hand he may not relish the idea of you producing your own trees for he has his livelihood to consider! So if you cannot obtain these rootstocks then you could stick to those which can be raised from seeds.

The following rootstocks are easily seed raised: *Acer platanoides*, *Acer pseudoplatanus*, *Crataegus monogyna*, *Fagus sylvatica*, *Laburnum anagyroides*, *Malus sylvestris*, *Prunus avium*, *Robinia pseudoacacia*, *Sorbus aria*, *Sorbus aucuparia*. For further details refer to Chapter 4, on Raising Trees and Shrubs from Seeds.

It is usual to graft rootstocks which are two years old, so that they are of a reasonable thickness. Plant out one-year-old rootstocks in the autumn and allow them to grow for another growing season, so they will then be ready for grafting in the second February or March from planting. Ideally they should be planted out in a nursery bed or in a spare piece of ground, in rows about 1 m (3 ft) apart, with about 60 cm (2 ft) between the rootstocks in the rows. The reason for suggesting a nursery bed or spare plot is because it will take up to four years to produce a tree from grafting – a tree that is of a suitable size for planting in the garden.

Method of grafting Now we come to the method of whip-and-tongue grafting, carried out in the dormant season in February or March, just before trees start into growth.

First prepare the rootstock by cutting it down to within 10 cm (4 in.) of the ground. Then make an upward slanting cut, about 5 cm (2 in.) in length, at the top of the rootstock. Near the top of this cut make a short downward cut to form a tongue. This operation is clearly shown in Figure 26.

Now to the preparation of the scion. From the plant you wish to propagate, remove some strong well-ripened shoots which were pro-

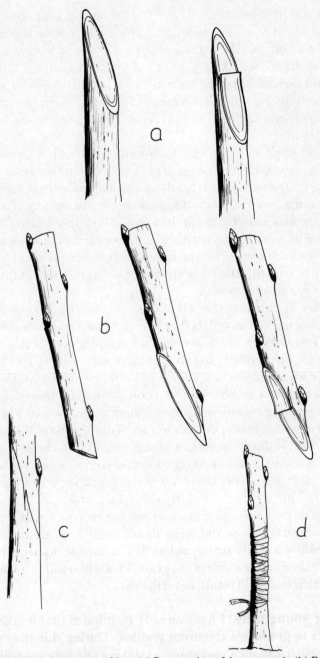

Figure 26 Whip-and-tongue grafting: (a) Preparation of the rootstock. (b) Preparation of the scion. (c) The scion joined to the rootstock – note the two tongues interlocking. (d) The graft tied in with raffia. This would be sealed to prevent entry of rain

duced in the previous year. These can be cut up into scions, which should be about four buds in length. The top of a scion should be cut just above a bud and the bottom about 2.5 cm (1 in.) below a bud. At the base of the scion make a 5 cm (2 in.) long downward slanting cut – you should then find that the bottom bud is situated midway along this cut, on the opposite side. Now a tongue should be cut in this long cut to match the tongue on the rootstock. Again this operation can be clearly seen in Figure 26.

The next stage is to join the scion to the rootstock, by pushing the tongue of the scion into the tongue of the rootstock until the scion is held in place. The graft should be tied firmly with raffia, as shown in the drawing, and the raffia covered with bituminous tree paint (e.g. Arbrex) to prevent rain from entering the grafted part. Also apply a dab of tree paint to the top of the scion to prevent diseases from entering the wound.

When the graft has united (an indication is when the scion is actively making new growth), the raffia should be slit at the back of the graft to allow for expansion of growth.

You may be thinking that all of this sounds fairly straightforward, but there are one or two points that I should make really clear to ensure success. To achieve a union between scion and rootstock it is essential that the cuts are perfectly flat so that there are no air spaces between. The latter often results in a failure to unite. I would strongly advise, therefore, that you practise your cuts on some spare material, such as fruit-tree prunings, until you can get them perfectly flat. For success you must also ensure that the green layer of tissue just under the bark of the scion is in intimate contact with the green layer under the bark of the rootstock. This green layer is exposed, of course, when you make the long cuts. It is only a very thin layer of tissue and is correctly called the cambium. This is where growth takes place and it is the cambium layers which produce new tissue, so uniting the stock and scion. The cambium layers of stock and scion should ideally match all round the graft to ensure a really strong union. If you cannot manage this, then match the layers on one side of the graft. This will result in a less-strong union but a tree should result, nevertheless.

Training young trees I have already mentioned that it takes about four years to produce a tree from grafting. During this time you will have to train it and such operations as staking and pruning will need to be carried out. I will deal with the training of standard trees as these are a popular form.

A standard tree has a clear stem (free of branches) to a height of about 2 m (6 ft) before it starts branching. Half standards are also popular in small gardens, especially for fruit trees, and these have a 1 m (3 ft) clear stem. Half standards are trained basically in the same way as standards. Of course in fruit growing trees can be trained to all manner of shapes, but this is a more specialized subject and best left to the expert. There are numerous books available which explain fruit-tree training techniques. There are some trees which do not need much in the way of training and these are left to grow quite naturally. Examples are fastigiate trees which are allowed to branch from as near the ground as possible, and also perhaps the weeping types could be allowed to develop naturally. Even with these, however, staking may be necessary in the early years to achieve an upright specimen.

To obtain a really straight trunk you should provide a stout bamboo cane about 2.5 m (8 ft) long (or a shorter one for half standards) in the year of grafting. You will find that the scion will make a number of shoots. Select the strongest of these (this is known as the leading shoot or the 'leader') and tie it to the cane with raffia, keeping it as straight as possible. The remaining shoots can be shortened by about half. Continue tying this leader as it grows (to the height required) to keep it perfectly straight.

You will find that lateral or side shoots are produced by the leading shoot – these should not be cut off immediately they appear as they help to encourage stem thickening. I prefer to wait until the autumn and then cut off the lower third of the lateral shoots completely. The remaining laterals can be shortened by a third to half their length. In the following autumn the middle third laterals can be removed and the remainder shortened. Finally in the third autumn the laterals at the top can be removed, so that you are left with a clear stem. The shoots which are produced beyond this form the head of branches. A good well-balanced tree should have about four initial branches as evenly spaced around the tree as possible. Any weak thin shoots can be removed to avoid a congested head or crown. By the end of the fourth year from grafting you should have a sizeable tree for setting out in the ornamental garden or fruit plot.

Saddle grafting

This method of grafting is used to propagate the large-flowered hybrid rhododendrons, such as the popular hardy hybrids. Admittedly some of these can be raised from cuttings but there are others, particularly many of the yellow- and red-flowered cultivars, which are difficult to grow from cuttings and so grafting is used to increase them.

Rootstocks – how to obtain and grow them The rootstock which is used for rhododendrons is the common or wild purple-flowered *Rhododendron ponticum*. The amateur gardener can raise this in two different ways. It can be propagated from semi-ripe cuttings in September or October: they should be rooted with strong bottom-heat – about 21°C (70°F). Root them in a propagating case or on a heated bench, covering the cuttings with polythene. I should add that this is not the easiest subject to root and therefore the amateur may find it better to raise rootstocks from seeds. Sow in heat in the spring. For further details see Chapter 4 which deals with raising shrubs from seeds.

Ideally two- or three-year-old rootstocks should be used for saddle grafting. These can be grown out of doors in pots and then taken into a heated greenhouse several weeks before grafting to start them into growth.

Method of grafting Saddle grafting is performed in March. It is important that the rootstock and scion are of the same diameter.

The rootstock should be cut down to within 5 cm (2 in.) of compost level. Then two cuts about 2.5 cm (1 in.) in length are made at the top of the rootstock to form it into a wedge shape. (See Figure 27.)

Now the scion is prepared. From the cultivar you wish to propagate select a strong previous year's shoot with a well-developed growth bud at the top. This should be cut to about 10 cm (4 in.) in length. Then two cuts about 2.5 cm (1 in.) in length are made in the base of the scion to form an inverted V shape. Now the scion is placed on top of the rootstock and if you have made the cuts really straight and flat you should find that the scion fits neatly over the wedge-shaped stock. As with whip-and-tongue grafting, ensure that the cambium layers of stock and scion match all round the graft. Now tie the scion tightly to the rootstock with raffia.

This graft needs heat in order to unite – preferably bottom-heat – so place in a propagating case or on a bench with soil-warming cables. It will take approximately six weeks for the graft to unite. Once there is a good union the raffia can be cut away. The young plants should now be gradually hardened: first they can be acclimatized in the normal green-

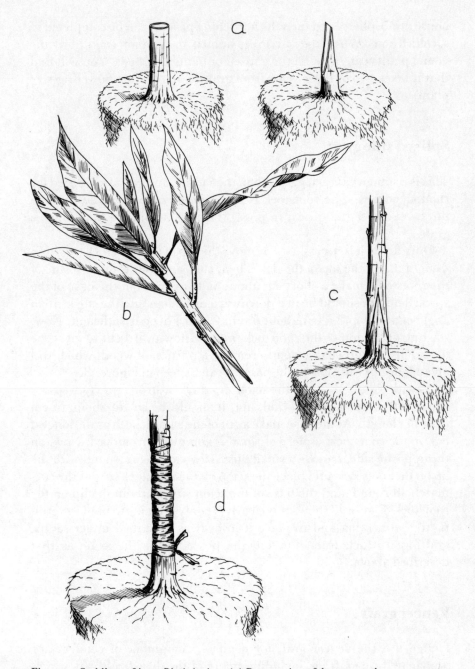

Figure 27 Saddle grafting a *Rhododendron*: (a) Preparation of the rootstock.
(b) Preparation of the scion – it is about 10 cm (4 in.) in length. (c) The scion sitting
neatly on top of the rootstock. (d) The completed graft tied in with raffia

house atmosphere, and then the hardening process can be completed in a cold frame. When they are accustomed to outdoor conditions the young plants can be set in the garden or in nursery beds. You will find that it takes two or three years from grafting to obtain a flowering-size plant.

Spliced side graft

This is a somewhat easier graft than the saddle, and can also be used for rhododendrons. The rootstock is again *Rhododendron ponticum*, two to three years old and grown in pots. March is a suitable time for this graft.

Once again it is necessary to remove the top of the rootstock to within 5 cm (2 in.) of the top of the pot. Then, choosing the straightest side of the rootstock, make a short cut about halfway up from the base of the stock. This cut should be in a downward direction, into the stock, at an angle of about 45°; a cut about 6 mm (¼ in.) deep is sufficient. Now, starting at the top of the rootstock, make a downward cut to meet the first. This should enable you to remove a portion of wood, which will give a long cut with a lip at the base, as can be seen in Figure 28.

As with saddle grafting, the scion should be a strong previous year's shoot with a good growth bud, and it should be cut to about 10 cm (4 in.) in length. At the base make a cut of the same length as the longest cut on the rootstock; a piece of wood is completely removed. Now, on the opposite side, remove a small piece of wood to correspond with the lip on the rootstock, and place the scion on the rootstock so that the cuts match all round, and the base of the scion sits neatly in the lip on the rootstock. Figure 28 shows this clearly. It only remains to bind the graft tightly with raffia and to place it in heat as described under saddle grafting. In fact, from now on, the procedure is the same as that described above.

Veneer graft

I often use the veneer graft for conifers. A number of conifers are propagated by grafting, including many of the dwarf cultivars which are often difficult to root from cuttings, plus the fact they are very slow to produce sizeable plants if they are on their own roots. Grafted plants

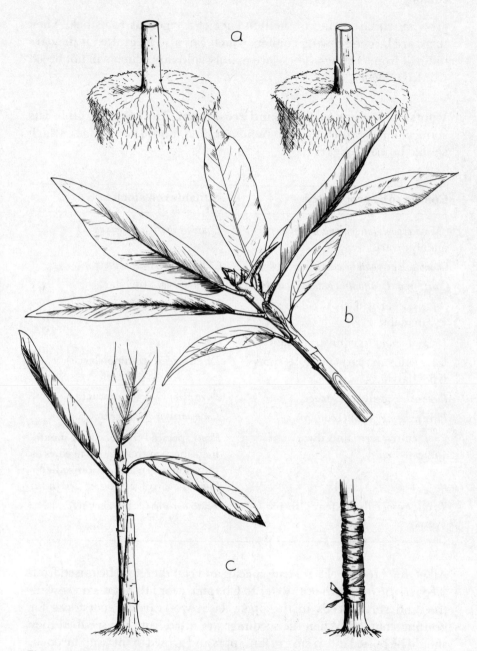

Figure 28 The spliced side graft, which can be used for *Rhododendron*: (a) Rootstock preparation. (b) Scion preparation, this being about 10 cm (4 in.) long. (c) The scion positioned on the rootstock: the base sits neatly in the lip on the rootstock. The graft is bound tightly with raffia

grow much faster due to the influence of a vigorous rootstock. Then there are larger-growing conifers which can also be grafted if they are difficult from cuttings. These are mainly cultivars which will not breed true from seeds.

Rootstocks – how to obtain and grow them The following table lists some conifers which are commonly grafted, and the rootstock which should be used for each.

Conifer to be propagated	Suitable rootstock
Chamaecyparis lawsoniana cultivars, mainly dwarfs	*Chamaecyparis lawsoniana*
Chamaecyparis obtusa cultivars	*Chamaecyparis lawsoniana*
Cedrus, e.g. *C. atlantica glauca* (cedars)	*Cedrus deodara* (deodar)
Cupressus, particularly *C. glabra* 'Pyramidalis'	*Cupressus macrocarpa*
Juniperus chinensis cultivars (junipers)	*Juniperus virginiana*
Larix cultivars, particularly weeping types (larches)	*Larix decidua* (common larch)
Picea abies cultivars (spruce)	*Picea abies* (Norway spruce)
Picea pungens glauca (blue spruce)	*Picea abies* or *P. pungens*
Pinus cultivars, mainly dwarf types (pines)	*Pinus* species – the rootstock should have the same number of needles as the scion; e.g. use the two-needled *Pinus sylvestris* for two-needled pines
Taxus, especially golden cultivars (yews)	*Taxus baccata* (common yew)

All of these rootstocks are true species and can therefore be raised from seeds in the open ground. Refer to Chapter 4 for full details on sowing tree and shrub seeds in the open. Use two-year-old rootstocks for grafting conifers. When the seedlings are lifted from the seedbed they should be potted into 9 cm (3½ in.) pots and grown in these in the open. A few weeks before grafting, which is carried out in March, the rootstocks should be taken into a heated greenhouse to start them into growth.

Method of grafting With the veneer graft we leave the top on the rootstock. Graft as low as possible – very near to compost level. To prepare the stock a vertical cut is made about 2.5 cm (1 in.) in length, taking the knife just under the bark. As will be seen in Figure 29, this will result in a flap of bark, with the wood beneath exposed. On no account make this cut too deep otherwise it will probably be wider than the prepared scion, and therefore the cambium layers will not match all round.

To prepare the scion, choose some previous year's shoots of the cultivar to be propagated and make them no longer than 10 cm (4 in.) in length. The length will depend on the amount of growth made by the conifer – many of the dwarf slow-growing kinds may only make about 5 cm (2 in.) of growth in a year and so the scions of these will of course be shorter than the maximum length suggested above. The base of the scion is cut into a thin wedge shape, by making two cuts opposite each other. The length of these cuts is the same as the cut on the rootstock.

Now the base of the scion is pushed down behind the flap of bark on the rootstock, ensuring the cambium layers match all round. If they do not, then match on one side only. It only remains to tie the scion with some thin raffia. (Nurserymen use 6 mm (½ in.) wide clear polythene tape but I doubt if the amateur could obtain this.)

The grafts are now placed in a propagating case with bottom-heat or on a bench with soil-warming cables. They will take approximately six weeks to unite. Once there is a good union cut away the tying material. The top of the rootstock is now cut off just above the grafted area. Then gradually harden the young plants as described under saddle grafting. Once fully acclimatized to outdoor conditions the young conifers can be planted out.

Spliced side veneer graft

As with the veneer graft, the spliced side veneer is also most successful in heated conditions. It is a useful graft for a number of trees and shrubs, mainly cultivars which cannot be raised from cuttings or which do not breed true from seeds.

Rootstocks – how to obtain and grow them The following table lists some trees and shrubs which are commonly grafted and the rootstock which should be used for each.

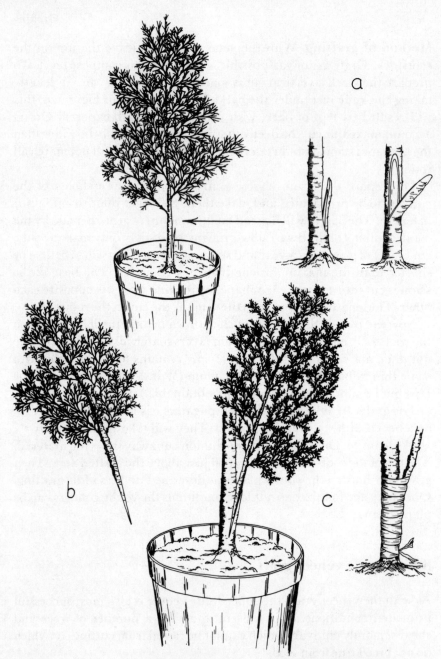

Figure 29 The veneer graft, useful for propagating conifers; this is a cultivar of *Chamaecyparis lawsoniana*: (a) Rootstock preparation, the cut resulting in a flap of bark. (b) Scion preparation, this being about one-third the height of the rootstock. The base is cut into a wedge shape. (c) The base of the scion is pushed down behind the flap of bark on the rootstock and is tied in with some thin raffia

Plant to be propagated	Suitable rootstock
Acer japonicum cultivars (Japanese maples)	*Acer japonicum*
Acer palmatum cultivars (Japanese maples)	*Acer palmatum*
Betula cultivars (birches)	*Betula pendula* (common silver birch)
Fagus sylvatica cultivars (beeches)	*Fagus sylvatica* (common beech)
Syringa cultivars (lilacs)	*Ligustrum vulgare* (common privet)

Again all of these rootstocks can be raised from seeds as described in Chapter 4. Two-year-old rootstocks in 9 cm (3½ in.) pots are needed for grafting and these should be taken into a heated greenhouse a few weeks before grafting is due to start, in March.

Method of grafting This graft is similar to the spliced side graft, except that the top of the rootstock is not removed. First comes the preparation of the rootstock. Graft as low as possible. Make a short cut near the base of the rootstock, in a downward direction into the stock, and at an angle of about 45°. The cut should be approximately 6 mm (¼ in.) deep. Now make a long downward cut to meet this short cut, the length being about 2.5 cm (1 in.). This should result in a piece of wood being removed, so that you have a long cut with a lip at the base, as can be seen in Figure 30.

Scion material should be previous year's wood cut to about 10 cm (4 in.) in length. At the base of the scion remove a strip of wood the same length as the longest cut in the rootstock. On the opposite side to this cut remove a small piece of wood to correspond with the lip on the rootstock. Now position the scion on the rootstock so that the base fits neatly into the lip of the rootstock. Ensure that the cuts of both scion and rootstock match all round. Finally bind the graft tightly with raffia and place in a propagating case or on a heated bench in the greenhouse.

After about six weeks the grafts should have united and then the raffia can be removed. The top of the rootstock should also be cut off just above the grafted area. Then gradually harden the young plants as described under saddle grafting, ready for planting out of doors.

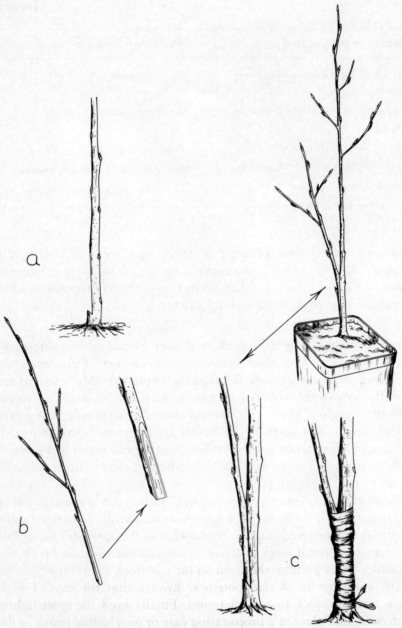

Figure 30 The spliced side veneer graft, useful for a number of trees and shrubs. Here, birch (*Betula*) is being grafted: (a) Rootstock preparation: a short cut near the base followed by a longer cut to meet it, resulting in a piece of wood being removed. (b) Scion preparation. The scion is about 10 cm (4 in.) long. (c) The scion positioned on the rootstock – the base fits neatly into the lip on the rootstock. Bind the graft tightly with thin raffia

9 Budding

Budding is really a form of grafting. As discussed in the previous chapter, a rootstock is used to form the root system of the new plant but, instead of grafting a piece of shoot or stem on to it, a growth-bud of the plant to be propagated is inserted in the stock. This bud unites with the tissue of the rootstock and then proceeds to grow to form the top part of the new plant. I have already discussed the reasons for using a rootstock for certain plants and the same comments apply when budding. Again you must use a rootstock which is closely related to the plant to be budded.

Budding is the usual method of propagating roses (*Rosa*) of all kinds – the bush types like hybrid teas and floribundas; the shrub roses; the climbers; and the ramblers. Most roses need the influence of a vigorous rootstock to grow and flower well, but there are some kinds which will grow well on their own roots, especially the ramblers, and so these could be propagated from cuttings, if desired.

Trees, both the fruiting and ornamental kinds, can also be propagated by budding as an alternative to grafting. Those discussed in Chapter 8 could be budded instead of grafted by the whip-and-tongue method, using the same rootstocks, of course. Many amateurs may in fact find budding easier and more successful than whip-and-tongue grafting.

Budding roses

Let us first deal with rose budding, which is done in the summer at any time between June and early September. One of the main problems for the amateur is obtaining rootstocks and so this will be dealt with in detail below.

Rootstocks – how to obtain and grow them The traditional rootstock for roses is the wild briar, *Rosa canina*, also known as the dog rose. This is found growing wild in hedgerows throughout the country. The problem with this species when used as a rootstock is that it suckers very freely – that is, it throws up many shoots from the root area. These shoots or suckers, if not regularly pulled or cut out, will compete with the cultivar which was budded on to the rootstock and eventually will take over. Another problem, perhaps more important to the commercial rose producer, is that it is rather variable in habit. *R. canina* is raised from seeds and the resultant seedlings lack uniformity in size and habit of growth. Nevertheless, the amateur will probably have to use *R. canina* as a rootstock for it is the easiest one to obtain. It is raised from seeds, after being stratified for about one year to eighteen months. Sowing is carried out in the open ground. Full details of stratification and sowing will be found in Chapter 4 under Trees and Shrubs from Seeds.

The one-year-old seedling rootstocks should be planted out in a nursery bed in October or November, spacing them about 30 cm (12 in.) apart in rows about 1 m (3 ft) apart. The rootstocks will then be ready for budding in the following summer. There is a special technique for planting rose rootstocks: the 'neck' of the rootstock (that is, the length of stem between the root system and the shoots) should remain above soil level. When the rootstocks have been planted soil should be drawn up around this neck to cover it, to keep it soft and succulent. A draw hoe can be used for this purpose and the soil should be ridged up along the row. When it is time for budding the soil should be pulled away to expose the neck. The bud is inserted in this part of the rootstock and you will find that the neck is very soft and easy to work if it has been kept cool and moist. If the neck is exposed between planting and budding it will become hard and woody and therefore much more difficult to insert the bud.

Well, so much for *Rosa canina* rootstocks; although I think it is worth mentioning that commercial rose producers now use special selections of this rootstock, instead of the straightforward species. The selections are generally less prone to suckering and the seedlings are much more uniform. There are many named selections, like the popular *R. canina* 'Pfander' and *R. canina* 'Inermis'. If you can persuade a kindly nurseryman to let you have a few of these then you will find them superior to *R. canina* itself. Another rootstock which is very popular with commercial producers is *R. dumetorum* 'Laxa', which is virtually thorn- less. I have found that this one can be raised from hardwood cuttings in the open ground. But, again, the amateur may have difficulty in obtaining this rootstock.

During the past year or so I have seen at least one firm offering rose-budding kits. So far as I know this is an innovation and could well solve the amateur's problem of obtaining rootstocks. The kit I saw was publicized in the gardening press and included rootstocks ready for planting, a budding knife and budding patches (which are used for holding in the bud once it has been inserted in the rootstock).

If you want to produce your own standard roses – a standard rose is like a bush rose on a clear straight stem – then the rootstock you should use is *Rosa rugosa* 'Hollandica'. This is a very prickly rose but it does produce good strong stems. It can be raised from hardwood cuttings rooted in the open ground. Disbud cuttings below ground level to prevent suckering. This method is discussed in Chapter 6. Once rooted a single stem should be trained up a stout bamboo cane to a height of at least 1–1.2 m (3–4 ft). It should be tied regularly with raffia to keep it perfectly straight. All side shoots should be rubbed out as soon as they appear. Shoots can be allowed to grow at the top of the stem once the desired height has been reached. Generally the stems are nearly two years old when they are budded.

Method of budding The time to bud roses is between June and early September; then the bark of the rootstock can be easily lifted when it is cut. The bark often does not lift easily if soil conditions are dry, so watering a few days prior to budding may be necessary. Just prior to budding the rootstocks the soil which was ridged up around the necks should be drawn away to expose them, as described above.

Let us first of all look at the method of obtaining buds to insert in the rootstocks. From the rose cultivar to be propagated remove a few well-ripened current year's shoots just after they have flowered, ensuring they have plump dormant buds in the leaf axils. Remove the soft tips from these shoots.

Now cut off all the leaves, but leave the leaf stalks or petioles. These prepared shoots are known as budsticks. It is advisable to stand the budsticks in a bucket of water to prevent them becoming desiccated, especially if the weather is very hot.

Now to the preparation of the rootstock. A T-shaped cut is made in the neck of the rootstock as close as possible to the soil. The longest cut should be approximately 2.5–3.5 cm (1–1½ in.) in length and the horizontal cut at the top of this need be only about 6–12 mm (¼–½ in.).

The cuts should be only very shallow – just sufficient to cut right through the bark. Now the bark on each side of the longest cut should be lifted with the spatula at the end of the budding knife. (Or some budding knives have a spatula at the top of the blade.) As soon as the cuts have been made on a rootstock a bud should be inserted. So now let us turn to removing a bud from the budstick.

First insert the budding knife 12–18 mm (½–¾ in.) below a bud, draw the blade under the bud and bring out the knife again 12–18 mm (½–¾ in.) above it. This will remove the bud on a shield-shaped piece of bark. If you inserted the knife only very shallowly there should be only a thin sliver of wood behind the bud. This piece of wood is generally carefully removed before the bud is inserted, although commercial nurserymen may leave it if it is very thin. Just flick out the piece of wood with the knife, ensuring you do not pull out the back of the bud as well. An indication of this is a small hole behind the bud; if the back is not damaged then there should be a small protuberance.

Now, holding the bud by the petiole or leaf stalk, slip it down under the bark of the rootstock. If there is any surplus bark sticking out above the T cut then it should be neatly cut off. All that remains now is to tie in the bud firmly. The conventional way is to use raffia, leaving the actual bud exposed. A newer way to tie in buds is to use the rubber budding patches mentioned earlier in this chapter. These are simply thin rubber squares about 2.5 cm by 2.5 cm (1 in. by 1 in.), each with a wire clip. The patch is placed completely over the bud, the two ends are pulled tightly round to the back of the rootstock, and one end of the patch is secured to the other by means of the wire clip. The whole operation takes about five seconds at the most once you have done it a few times, and it is certainly a lot quicker than tying with raffia. Within a few weeks these budding patches rot and by this time the buds will have united with the rootstocks.

If you intend budding standard roses then the buds should be inserted in the top of the stem, at the height you require the branches to form the head. Generally three buds are inserted, as this will ensure a well-balanced head, and they should be arranged in a spiral formation.

Aftercare of budded roses Once budding is completed there is very little to do until the following late winter. In the meantime the budded rootstocks should be kept watered, if the soil starts to become dry, and it will also be desirable to keep weeds in check. Thick weed growth could smother the buds.

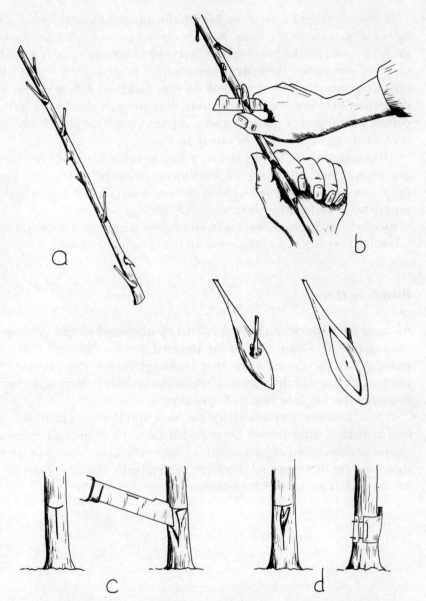

Figure 31 Rose budding: (a) A budstick of the cultivar to be budded – note the petioles have been left on and that there are plump buds in the axils. (b) Removing a bud on a shield-shaped piece of bark. The thin sliver of wood behind the bud should be carefully removed to expose the back of the bud – see right-hand drawing. (c) A T-shaped cut being made in the bark of the rootstock. Lift the bark on either side. (d) The bud is slipped down under the bark of the rootstock and is held in place with a rubber budding patch

Budding

In the following February or March the top of the rootstocks should be cut off just above the buds. Be very careful that you do not damage the buds, most of which will still be in the dormant stage at this period – they are sometimes difficult to see clearly. Sometimes the odd bud starts into growth in the summer following budding. This is the exception rather than the rule and such buds are known as 'shot' buds. When cutting off the top of the rootstocks any shot buds should also be cut back hard – to within a bud or two of the base.

Throughout the following summer the roses should make a fair bit of growth and you will even get a few flowers on the bushes. Again keep the plants well watered and weeded and you could also feed them with a proprietary rose fertilizer to ensure really good growth.

By the following autumn, fourteen to sixteen months from budding, you will have roses of a suitable size for planting in the garden.

Budding trees

All those trees which can be propagated by whip-and-tongue grafting – see Chapter 8 – can also be propagated by budding. The actual technique is the same as for rose budding, except that the bud is inserted higher – in the region of 10–20 cm (4–8 in.) above soil level – during the period June to early September.

When the rootstocks have been cut back and the buds have started into growth a cane should be provided for each plant and training should be carried out as mentioned in Chapter 8. There will be only one shoot or stem this time and this must be tied really straight to ensure a trunk which is not kinked or twisted.

10 Division of plants

Division, or splitting plants into a number of smaller portions, is a method of increase that even the most inexperienced propagator can attempt and achieve complete success. Many plants which form clumps, mats or carpets of growth, such as hardy perennials, alpines, aquatics, greenhouse plants, some shrubs and bulbous subjects, can be propagated by division. Let us now deal with each of these groups in turn.

Hardy perennials

Herbaceous plants and other hardy perennials will form large clumps or 'crowns' after several years at which time they should be lifted and split into smaller portions and replanted. Generally many young plants can be obtained from one large clump; often there is more than required for replanting and so the surplus can be given away to friends and neighbours.

If you allow perennials to form very large clumps they will start to decline in vigour and flowering will be poor. Therefore it is the usual practice to lift and divide such plants every three or four years to keep stock young and vigorous. I prefer to lift and divide in the period March to mid April rather than the autumn, just as the plants are starting into growth after their winter rest. At this time the weather and soil are warmer, and the ground is often dry; this ensures swift establishment of the divisions – they will quickly root into the soil. In the autumn, when the soil is cooling down and becoming wet, the plants will not make root growth and there is a likelihood that roots will rot and the plants die.

The method of division The clumps, mats or carpets should be lifted with a garden fork. Insert the fork to its full depth all round the plant, prising it back each time until the plant is loose and can be easily lifted. Remove as much soil as possible without damaging the roots.

The easiest way of splitting large clumps is to thrust two forks back to back right through the centre and to pull the handles apart. This will result in two divisions, which can be split further by the same method. To reduce the divisions still more, you may now be able to pull them apart with your hands, or by using two hand forks placed back to back. You may also find an old knife useful for cutting through very tough crowns, but avoid damaging too many roots.

The next consideration is which parts of the plant to retain for replanting and which to discard, remembering that each division should consist of a number of growth buds or shoots plus a good portion of fibrous roots. Always discard the centre part of each plant as this is the oldest and therefore will be declining in vigour. Retain the outer parts of a clump as these are young and vigorous. Now what about the size of each division? Generally speaking divisions which sit comfortably in the palm of the hand are about the right size for replanting, to ensure a good flower display the same year.

Divisions should be replanted with the minimum delay to prevent the roots from drying out. Do not forget to keep them watered if the weather is dry following planting, to ensure quick establishment.

Having given a general time for the division of perennials, I must now add that the early-flowering kinds are best divided immediately after flowering. Examples include the spring-flowering *Doronicum*, *Epimedium*, German or bearded *Iris*, *Primula denticulata*, other primulas of the primrose type, *Pulmonaria* and *Pyrethrum*. With *Iris* the method of division is slightly different – each division should consist of a portion of rhizome or swollen stem with fibrous roots attached, plus a fan of leaves. When replanting these, ensure that they are made really firm and the rhizomes are only lightly covered with soil. (See Figure 33.)

Also remember that there are some perennials which are best not lifted and divided as they resent root disturbance. These include *Anemone × hybrida* (the Japanese type), *Alstroemeria, Eryngium, Echinops, Helleborus, Paeonia* and *Romneya coulteri*. Propagate these by other methods if possible – see the tables for further information.

You will find that there are some perennials which cannot be divided because they do not form clumps but produce a single crown or maybe a taproot, or several thick fleshy roots, instead of a fibrous root system.

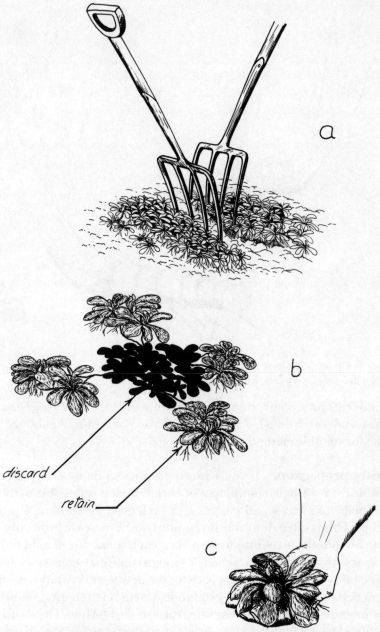

discard

retain

Figure 32 Dividing a herbaceous plant: (a) Large clumps can be split with the aid of two garden forks thrust through the centre and the handles pulled apart. (b) Discard the centre part of each clump and retain the young vigorous outer parts. (c) Divisions which sit comfortably in the palm of a hand are about the right size for replanting

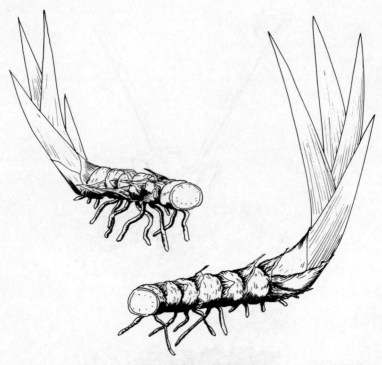

Figure 33 Divisions of German or bearded *Iris*. Each consists of a portion of rhizome or swollen stem, some roots and a fan of leaves

In this category some common examples include *Anchusa*, *Dianthus* (pinks and carnations), *Lupinus* and *Papaver orientale*. Again see the tables for suitable methods of increase.

Thrifty propagation If you wish to produce as many new plants as possible from a single clump, mat or carpet then it is possible to divide perennials into very small pieces, with each piece resembling a rooted cutting. This is often desirable if you want quick increase from only one plant. So split the plant into small pieces, each consisting of a single bud or shoot with a few roots attached, although some may consist of several buds or shoots. I like to do this early in the new year. With this method it is necessary to encourage a good root system on each division before planting out, which is done in late April or early May. The divisions can therefore be planted close together in deep seed boxes or tomato trays of cutting compost (equal parts peat and coarse sand). Stand the boxes in a sheltered spot out of doors, and water whenever necessary. By planting time the divisions will be well rooted, and they can then be planted, either in a nursery bed for a season, or direct in the ornamental bed or border. By the following autumn they will have made fairly large plants. In fact, most subjects will flower in the summer or autumn of the same year.

This method of propagation is often practised by nurserymen who want to produce as many plants as possible in the shortest possible time, but it is an equally useful method for the amateur gardener who wants to quickly increase a stock of perhaps choice or expensive perennials.

Dicing rhizomes of bergenias Bergenias are evergreen perennials which flower in the spring; in recent years many new cultivars have been introduced. At the present time these are rather on the expensive side and many people are unable, or do not desire, to buy them in quantity – so you may be interested to know of a method of propagation which quickly increases them.

These plants produce thick fleshy rhizomes or swollen stems at soil level and it is from these that new plants are raised. When bergenias have been established for a few years there should be a good number of rhizomes on each plant.

In January the plants should be lifted and the rhizomes cut off. If you do not cut these back too hard, and the plant still has some fibrous roots, then the parent plant can be replanted and it should re-establish.

Now thoroughly wash the rhizomes to remove all trace of soil. Using a sharp knife cut the rhizomes into sections about 2.5–3.5 cm (1–1½ in.) in length. The rhizomes are compressed stems and there should be many dormant growth buds on each, so each section of rhizome should have at least one bud. The buds are generally small and rather difficult to see, but it is possible if you search for them. The rhizome sections are then lightly dusted with Captan fungicide to help prevent rotting – shake them about in a polythene bag which contains a little Captan dust.

It is necessary to root the diced rhizomes in a sterile medium. I generally use Perlite, but Vermiculite could also be used. Fill a seed box with the rooting medium and then insert the rhizome sections horizontally, with the bud uppermost if you have managed to find it. Simply press the sections into the rooting medium to about half their depth. Water them thoroughly, using a can with a rose.

Strong bottom-heat is required for rooting, so place the box of rhizomes in a propagating case or on a warm bench in the greenhouse. A temperature of about 21 °C (70°F) is desirable at the base.

You will find that they very quickly make new shoots and leaves but this does not mean that they have formed an adequate root system – in fact they generally produce top growth before roots – but after several weeks they will be well rooted in the box. Then they can be lifted and potted into 7.5 cm (3 in.) pots of John Innes potting compost No. 1. Return them to the greenhouse, and once they are well established in their pots transfer the young plants to a cold frame to gradually harden them. The young bergenias can be planted in the garden or in nursery beds in May, and by the following autumn they will have made reasonably large plants.

Grasses and ferns

In recent years ornamental grasses have become very popular, and I think it safe to say that ferns are regaining the popularity which they enjoyed in the Victorian era. These two groups of plants are clump formers and therefore increase by division is possible.

Again the best time of the year to lift and divide ferns and grasses is in March to mid April, just as they are starting into growth. Very often the crowns of ferns and also of some grasses, especially pampas grass (*Cortaderia*) and zebra grass (*Miscanthus sinensis* 'Zebrinus'), are very tough and you may need the help of an old knife to cut through them. *Cortaderia*, incidentally, is difficult to re-establish after division and some losses may occur, so be prepared for this.

The majority of ferns and grasses benefit from division about every three or four years, not only to prevent the clumps from becoming too large but also to ensure that the plants remain young and vigorous. In general use the method of division recommended under hardy perennials, and replant divisions about the size of the palm of your hand.

Alpines

Division is a suitable method of increasing a wide range of alpines or rock plants and indeed nurserymen often use this method and are able to produce substantial plants in 9 cm (3½ in.) pots within twelve months. It is the mat-forming and carpeting types which are suited to this method of propagation, as well as those which form offsets or young plants around themselves. Alpines which form cushions or rosettes of growth may also be easily divided.

I find the best time for alpine division is in the early spring just as the plants are starting into growth, as then the divisions establish very quickly. Some people, however, prefer to split plants in late summer or early autumn.

Dividing mats and carpets Plants which come in this category include *Saxifraga* of various species and cultivars, *Raoulia* and thymes (*Thymus*), plus many other well-known genera. Many of the saxifrages also form cushions or rosettes of growth. Plants of this habit are lifted carefully with a fork and they can then be gently pulled apart into pieces about 3.5–5 cm (1½–2 in.) in diameter. Ensure that each division has some roots.

Removing offsets Certain alpines have the habit of vegetatively propagating themselves by means of offsets, which are simply young plants produced around the parent plant. Typical examples of alpines with this characteristic are houseleeks (*Sempervivum*) and *Androsace sarmentosa*. Generally you can remove these offsets without disturbing the parent plants. Ease them out of the soil with a small hand fork and pull them away from the parent plant – they are usually attached to short stems.

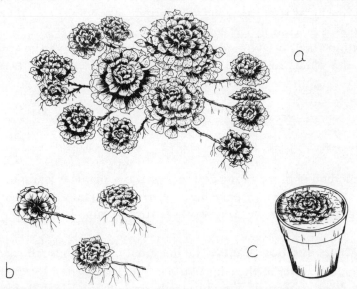

Figure 34 *Sempervivum* offsets: (a) Sempervivums produce young plants or offsets on short stems. (b) Rooted offsets can be carefully removed without disturbing the parent plant and (c) potted individually into small pots

131

Growing on divisions Most people would probably prefer to replant their divisions in the rock garden, and of course this is quite satisfactory. But if you wish to grow the divisions into larger plants before planting, pot them into 9 cm (3½ in.) pots. The compost I generally use for alpines is a modified John Innes potting compost No. 1. To a volume of JIP1 I add about half the volume again of grit or coarse sand to ensure a really open, well-drained, well-aerated compost which these plants enjoy. Many alpines are not the easiest subjects to pot as, being rather flattish in habit, you invariably get compost all over the foliage – so pot with care and try to avoid this happening as much as possible. Compost is extremely difficult to remove from rosettes, mats and carpets of growth. I normally place one hand over the entire plant while filling the pot with soil with the other. Pot reasonably firmly and then water in with a rosed can.

The pots of alpine divisions can then be placed in a sunny sheltered spot to grow. Remember these small pots will dry out rapidly in warm weather so keep a regular eye on watering. If the pots are plunged to their rims in well-weathered ashes this will prevent very rapid drying.

Plants in pots are best overwintered in cold frames to prevent them becoming too wet. However, they should not be grown 'soft', so prop up the frame lights to ensure good ventilation – it is not the cold they mind so much as excessively wet conditions.

Within about twelve months from division, the pots will be well filled with the plants and then they can be planted out. The best time for planting alpines is in the spring, if possible.

Aquatics

The method of propagating aquatics or water plants is again division. After all, aquatics are only hardy perennial plants and often form clumps of growth in the same way as border perennials.

Marginals and bog plants The marginal plants (those that grow in the shallow water at the edge of a pool, such as reeds (*Typha*), rushes (*Juncus*) and water *Iris*), and the bog plants (subjects which thrive in the moist soil at the edge of a pool, like bog *Primula* and *Lysichitum*) are all divided in exactly the same way as recommended for hardy border perennials. The only difference is the time of year, for water and bog plants are lifted and divided in the period April to June. When you are dividing these plants take the opportunity of removing any dead foliage, which is inclined to foul the water. Again all of these subjects will benefit from being split every three to four years. It is generally a good idea to wash the roots free of soil and mud before dividing the plants as then the operation is much easier and you will be able to see exactly what you are doing.

Submerged oxygenating plants The submerged oxygenating plants or water weeds, which are essential for a well-balanced pool, are generally very vigorous in habit and need lifting and reducing in quantity every two to three years. When removed from the pool they will present a tangled mass of growth. This growth should be gently pulled apart into small bundles which can be held together at the base with a small elastic band. Then the appropriate number of bundles for the size of the pool should be replanted. You will have a surplus of plants which can either be given away or discarded. It does not matter if some of the stems which are replanted do not have any roots for they will very quickly produce new ones if divided from April to June.

Waterlilies The waterlilies or *Nymphaea* are also divided in April to June. They produce their top growth from tubers, which should be lifted and thoroughly washed before dividing. These tubers are cut into pieces to make new plants and a sharp knife will be needed as often they are very tough. Each piece of tuber must contain a substantial crown with some growth buds or 'eyes' as it is from these that new shoots and leaves are produced. Any very long roots can be cut back to make replanting easier. From a large tuber you should be able to obtain many new plants. Waterlilies should be lifted and divided every three to four years to prevent overcrowding. (See Figure 35.)

Do not propagate weeds One final thought in the propagation of aquatics – make sure you remove all traces of blanket weed from the divisions before replanting as this is a great problem in a pool and difficult to eradicate among established plants. It presents itself as fine green filaments and every scrap should be removed from divisions, to prevent this weed from increasing again.

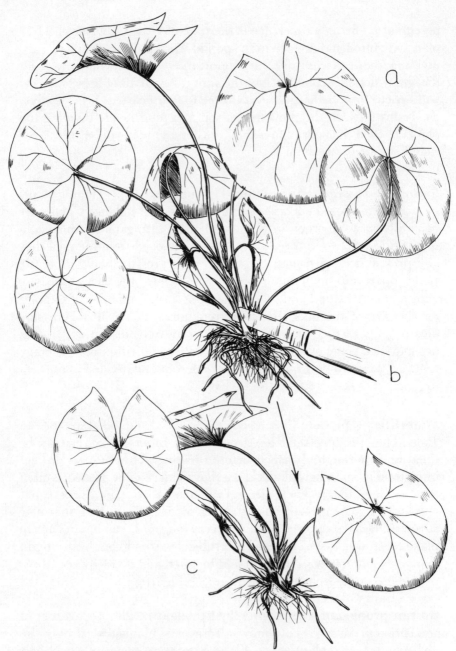

Figure 35 Dividing a waterlily (*Nymphaea*): (a) This is a miniature waterlily with a small tuber. (b) The tuber can be cut into pieces to make new plants. (c) Each division should consist of a piece of tuber plus some growth buds (and maybe a few leaves) and some roots

Greenhouse plants

Many greenhouse plants or pot plants form clumps and so division in the spring is a suitable method of propagation. Often these plants become very large after a number of years and they need to be reduced in size so that they fit into reasonable-sized pots. Among the clump formers are *Aspidistra*, bromeliads, many cacti and succulents, *Chlorophytum*, many of the ferns, a wide range of orchids, *Sansevieria* and *Saintpaulia*.

Plants should be removed from their pots and as much soil as possible teased away from the roots to make division easy and to avoid too much root damage during the operation. Then split the plants into a number of smaller portions, either by carefully pulling them apart or with the use of a knife if the crowns happen to be on the tough side. Try to ensure that each division has adequate top growth plus a good portion of roots.

Now choose suitable-sized pots for the divisions and repot them, using fresh potting compost. I generally prefer to water newly potted plants to settle the compost around them. If possible keep the newly potted divisions in a warm place for a week or two to encourage them to quickly put forth new roots. Shading from the strongest sun will also help them to become quickly established. When well established, provide normal temperatures according to the species.

Orchids I mentioned in general terms above that orchids can be increased by division but as this is a rather specialized group of plants, more detail is needed. Division can be carried out when plants have two or more new growths, but do make sure that sufficient pseudobulbs remain after division. Pseudobulbs are the large green bulb-like growths produced by orchids and the term simply means 'false bulbs'. There are some orchids which can be increased by detaching offshoots which are produced by the parent plants. These should be potted off into small pots. *Vanda* species often produce offshoots.

Cymbidiums are often propagated by backbulbs. These are the older, leafless pseudobulbs behind the leading pseudobulbs. They should be carefully removed (generally they have no roots) and potted individually into suitable sized pots.

This is a convenient place to mention propagation of orchids by plantlets. Some *Dendrobium* species can be increased by means of plantlets formed on the parent plants. Carefully remove the plantlets and pot them.

Shrubs

Certain shrubs have what is known as a suckering habit – that is, they spread outwards by means of shoots which originate from the root area. Very often thickets of shoots can be found all around the parent plants and in some of the very vigorous subjects this means of increase often becomes a nuisance as the plants soon spread out of their allotted space. Very often gardeners have to pull up much of this growth (or suckers) to contain the shrubs. But such material is useful for propagation as the suckers are generally well rooted and can be replanted elsewhere to provide new plants. The best time of the year to remove rooted suckers from shrubs is in the dormant or resting season – between November and March.

A word of warning, however, before discussing the method in more detail. Shrubs or trees which have been propagated by budding or grafting often throw up suckers from the roots of the rootstock on which the variety or cultivar was budded or grafted. On no account use these for producing new plants for the resultant plants will have the characteristics of the rootstock and not of the variety or cultivar. The rootstocks are generally the common counterparts of the variety or cultivar and more often than not they are uninteresting plants. Suckers of this type should be dug or pulled up and discarded otherwise they will eventually dominate the variety or cultivar.

Removing and growing suckers Some common examples of plants which sucker freely include *Clerodendron*, shrubby *Cornus*, *Rhus typhina* and other species, raspberries (*Rubus*), *Symphoricarpos* and *Yucca*.

To remove suckers for transplanting carefully lift them with a garden fork and shake them free of soil, then sever them from the parent plant with a pair of secateurs. Aim to lift suckers with as much of the fibrous root as possible, to ensure good re-establishment.

Very often, depending on the subject, you will have quite large rooted portions to replant and so fairly large new plants will be obtained in a short time. Do not let the rooted portions dry out between lifting and replanting; replant as soon as possible after lifting, firming the pieces well with your heel.

Bulbs, corms and tubers

Plants which grow from bulbs, corms and tubers can often be increased by various forms of division. Let me first deal with the true bulbs, such as hyacinths (*Hyacinthus*), daffodils (*Narcissus*) and tulips (*Tulipa*).

Bulblets These are simply small bulbs which are produced around the base of a parent bulb. Dig up any typical bulbous subject, for example a daffodil or a tulip, after it has been established for a few years, and you will find quite a colony of young bulbs or bulblets. This is a completely natural method of increase, of course, and, depending on the subject, bulblets will grow to normal flowering size in one to three years.

To remove bulblets dig up the parent bulbs only when the foliage has completely died down which, for the majority of subjects, will be late spring or during the summer. Do not dig up bulbs when the leaves are still green for they will be manufacturing foodstuffs which ensure an increase in the size of the bulbs. Having said this, however, I must now add that *Galanthus* or snowdrops are best lifted immediately after flowering: separate the bulblets and replant immediately.

Once the bulbs of other subjects have been lifted, pull away the bulblets and then dry them thoroughly if you wish to store them in a cool dry place until the early autumn. Or you could detach and plant them at the appropriate planting season.

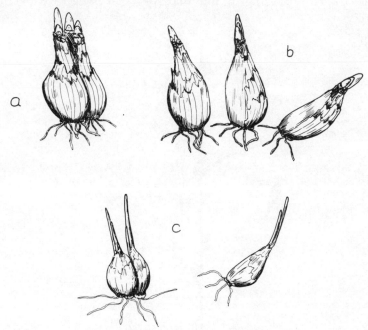

Figure 36 Removing bulblets: (a) A parent daffodil bulb with bulblets around it. (b) The bulblets separated from the parent bulb. (c) A snowdrop (*Galanthus*) bulb with a bulblet attached. The right-hand drawing shows the bulblet removed

When planting the bulblets in the early autumn do ensure that they are inserted shallowly: they will not be fully grown and therefore do not need the planting depth recommended for full-size bulbs. They can either be planted in a nursery bed to grow to flowering size, or you may prefer to plant them in groups in the garden.

Bulbils The propagation of bulbous plants by bulbils is not really a form of division in the recognized sense, although it is convenient to deal with the subject under the heading of bulbs.

Bulbils are small bulbs which are found on the stems of some plants and they are usually formed in the leaf-axils. Many lilies produce bulbils, such as *Lilium bulbiferum, L.* 'Enchantment', *L. maculatum, L. sargentiae, L. speciosum, L. sulphureum* and *L. tigrinum.* Various members of the onion family also have this natural method of vegetative propagation, such as tree onions, garlic and some of the ornamental species of *Allium*.

Bulbils will take from two to three years to form flowering-size bulbs if they are removed and grown on; the best time to gather them is in late summer. It is interesting to note that specialist lily growers actually

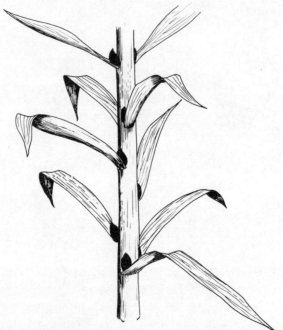

Figure 37 Bulbils or small bulbs on a stem of a lily (*Lilium*). These are gathered in late summer and can be stored in peat for the winter, to be planted in the spring

encourage the formation of bulbils by cutting back the flower stems to about half their length before they come into flower. This encourages a mass of bulbils to form in the leaf axils. The amateur gardener could try this if he or she can bear to sacrifice a few flower stems!

The bulbils should be dried if they are at all wet and then they should be stored for the winter in dry peat. They could, for example, be stored in seed boxes. Keep them in a cool, dry, frost-proof place over the winter.

In the spring the bulbils should be planted in rows in a nursery bed to grow to flowering size. As the bulbils are only very tiny, they are virtually 'sown', like seeds, in 5 cm (2 in.) deep drills taken out with a draw hoe. When they have reached flowering size the bulbs can then be planted in the ornamental garden, lifting them when the foliage has died down.

Lily scales Lilies (*Lilium*) can be propagated by removing scales from the bulbs and inducing these to root and form new plants. Before discussing this method in detail it will be as well to clarify the meaning of the term scales. Bulbs, lilies included, are composed of scales, which are the fleshy swollen bases of leaves. Their function is to store foods and water during the resting period, to keep the bulb alive until new roots are formed when the bulb has started into growth again. A bulb consists of many layers of scales which are joined to a plate-like structure at the base; this is in fact a modified stem.

Lilies are generally propagated from scales, which is really a form of division, when flowering is over – this is generally in late summer. Propagation by this method can also be undertaken in the spring.

First lift a few bulbs and then break off a few of the scales, ensuring that they come away as close to the base of the bulb as possible. The bulbs can then be replanted.

The scales are inserted in a seed tray containing a mixture of equal parts peat and coarse sand. Insert them in an upright position to about half their depth, ensuring that they are the right way up.

There are various places in which lily scales can be rooted. For quickest results use a propagating case or place them on a bench in a heated greenhouse. Alternatively rooting will take place in a cold frame. Water sparingly to start with – keep the compost only slightly moist otherwise rotting may occur. Eventually the scales will root and a small bulb will be produced at the base of each. At this stage one can with advantage increase the amount of water given.

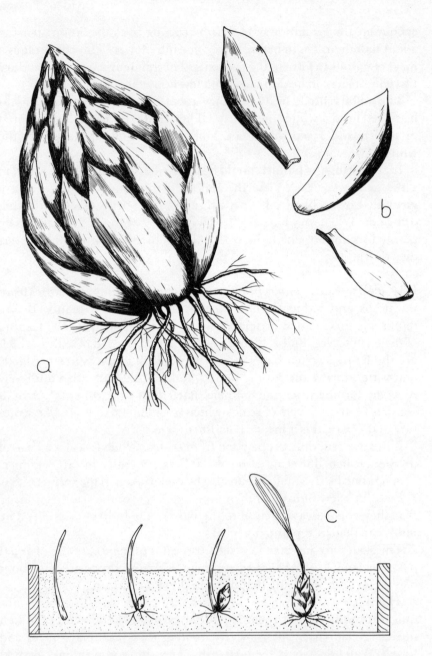

Figure 38 Propagating a lily (*Lilium*) from scales: (a) A mature lily bulb clearly showing the scales. (b) A few scales can be carefully pulled off such a bulb: they must come away as close to the base of the bulb as possible. (c) The stages in the development of a young lily bulb from a scale. The scales are inserted upright in a seed tray of compost

When well rooted, put the scales into individual small pots, using John Innes potting compost No. 1. The young lilies should be gradually hardened in cold frames before being planted outside, either in nursery beds or direct in the ornamental beds and borders.

Depending on the species or cultivars, it will take about two or three years for the young bulbs to reach flowering size. This is certainly an economical way to increase lilies, which are quite expensive to buy nowadays.

Cormlets Now we come on to the propagation of corms. It is subjects like *Crocus* and *Gladiolus* which grow from corms. These are somewhat different from bulbs, for corms are compact swollen stems; but again their function is food and water storage to keep the plant alive during the resting period.

Cormlets are small corms which are produced around the base of the parent corm – a natural method of vegetative reproduction. If the parent corms are lifted when the leaves have completely died down you will very often find a mass of these cormlets. These should be removed and stored in seed trays of dry peat during the winter, choosing a dry, cool but frost-proof place.

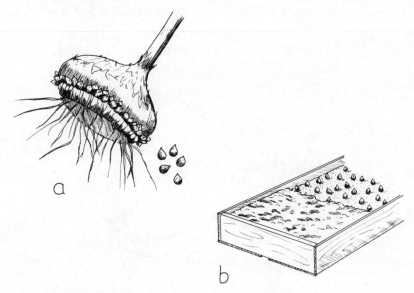

Figure 39 *Gladiolus* cormlets: (a) Cormlets are small corms produced around the base of the parent. (b) The cormlets can be removed, stored in trays of peat over the winter and planted in the spring out of doors

In the spring the cormlets are planted in rows in a nursery bed to grow to flowering size. Plant in drills 5 cm (2 in.) deep, taken out with a draw hoe. *Gladiolus* should be lifted in the autumn each year and stored dry in a frost-proof place, then planted out again in the spring. But hardier subjects like *Crocus* may be left in the ground over the winter. Once the corms have reached flowering size plant them in the garden; it will take about two to three years for young corms to come into flower.

Corms can be artificially induced to produce cormlets by making a few cuts across the base of the corms before planting. But generally speaking the parent corms will produce sufficient offspring without the necessity of this artificial stimulation.

Gladiolus corms can also be increased by division. The corms are cut in half before planting, ensuring that each portion has a dormant bud at the top, otherwise no leaves and stems will be produced, and that the base of each division contains part of the plate from which new roots will be produced. The cut surfaces of the divisions should be dusted with Captan or a similar fungicide to prevent rotting. Allow time for the cut surface to heal before planting.

Tubers The *Dahlia* is an example of a plant which forms tubers, and in this instance the tubers are swollen roots which act as food and water-

Figure 40 Dividing a *Dahlia*: (a) The dormant plant can be divided in the spring before planting. (b) Each division should consist of a piece of crown containing growth buds and at least one tuber

storage organs to keep the plant alive during its winter resting period. The shoots of a *Dahlia* arise from growth buds located in the 'crown' which is found just above the tuberous roots.

Dahlias eventually make massive clumps and so should be regularly divided prior to re-planting the dormant plants in the spring. Each portion or division should consist of a piece of crown containing a number of growth buds, plus at least one tuber. If tubers do not have a portion of crown they will not grow as they do not contain growth buds. You will find that some cultivars are easily divided: they can be pulled apart with the hands. But other cultivars are very tough as they form massive tubers tightly packed together, and in this instance you will need a sharp knife to divide the clump. Any cut surfaces should be lightly dusted with flowers of sulphur to prevent rotting.

Remember that dahlias are half-hardy perennials and must be lifted in the autumn, dried and stored in a cool but frost-proof place over the winter. The dormant tubers are then replanted in about mid April, covering the crowns with about 7.5–10 cm (3–4 in.) of soil.

11 Simple layering

Perhaps the easiest and most reliable method of vegetative propagation for the amateur gardener is simple layering. Basically this involves pegging down a stem or branch into the soil while it is still attached to the parent plant. Roots will be produced on the part of the stem in contact with the soil. The rooted part is later removed from the plant, ready rooted, for setting out elsewhere in the garden.

Layering is not an artificial method of propagation as it often occurs in the wild. The shoots of brambles (*Rubus*) will root into the soil at their tips, for instance, and I have even seen the branches of beech trees (*Fagus*) rooting into the soil where they sweep down to the ground. In the garden you may find other plants layering themselves, such as *Forsythia suspensa*, whose weeping branches often touch the soil, and low-growing or prostrate shrubs like the ground-covering cotoneasters, *Juniperus* and *Viburnum plicatum*.

Simple layering is used only for those trees and shrubs whose shoots or branches can easily be brought into contact with the ground. There is no point in forcing branches to soil level, with the possibility of breaking them in the process, as there is another method of layering known as air layering which could be used instead (see Chapter 12).

Simple layering can result in 100 per cent success and, even more attractive for the amateur, it enables plants which are difficult or impossible to increase from cuttings and other methods to be propagated. The offspring will be identical in every way to the parent plants.

No special equipment is needed for simple layering. All that is necessary are a few wire pegs and some short canes, plus soft garden twine or raffia. Peat and sand would also be useful. What you will need in abundance is patience, for it may take eighteen to twenty-four months for some subjects to form a substantial root system in the soil.

18 Preparing a seed bed for sowing hardy annuals. Here cultivated ground is being firmed thoroughly by treading

19 The firmed ground is now being raked down to a fine tilth for sowing

20 Drills for sowing the seeds are being taken out with the edge of a draw hoe

21 The seeds are being sown thinly and evenly in the drills

22 The seeds can be covered in various ways. Here the back of a rake is being used to gently draw soil back into the drills

23 Seedlings of hardy annuals which are well in need of thinning out

24 The same batch of seedlings thinned out, the remainder being evenly spaced with adequate room for development

25 The back of a fern frond showing spore cases or sori

26 The prothallus stage of fern propagation from spores. These have been pricked out into a seed tray to further develop

27 Young ferns ready for potting into individual pots

28 A newly potted young fern

29 A modern electrically heated propagating case, ideal for rooting cuttings or for germinating seeds

30 Part of a small mist-propagation unit, showing a mist head and (bottom left) an electronic leaf or moisture-sensitive device

31 *Juniperus* cuttings rooting in a cold frame

32 Cuttings of many plants can be rooted in pots on a windowsill indoors, especially if they are covered with a clear plastic 'bell glass'

33 Cuttings of some plants (like *Impatiens*, as shown here) can be rooted in water on a windowsill indoors

34 A small home nursery showing rows of conifers and other plants being grown to a larger size for setting out in the main part of the garden

On no account be tempted to lift shoots or branches until they have spent the requisite time in the soil, for a poorly rooted shoot will not transplant and establish very well and it may even die in the process.

Time of year to layer stems There is no strong reason why you should not peg down stems at any time of the year, but I prefer to choose a period when the plants are actively growing as rooting should then be quicker. April to August is, I find, a good period. Choose a time when the soil is not too wet – you should be able to break it down to a fine tilth. If you layer in the winter when plants are dormant, the stems will not start to root until the soil warms in the spring and the plants are starting into growth again. There is then the possibility that the part of the stem in the soil will rot during the winter because the wound that has to be made will take a long time to heal. So, if possible, layer when the plants are active.

Choosing suitable stems The stems which are pegged down are popularly known as 'layers'. They must be chosen with care. Young or old trees and shrubs can be layered; age does not affect this method of propagation provided there is some young wood to peg down into the soil. In fact, if you have a very old plant in your garden that you wish to replace, why not layer some branches to provide young plants before you grub out the old one?

Only peg down young wood – preferably wood formed in the current or previous season. Older wood may not root; or if it does it may take an unreasonably long time in doing so. On the other hand young wood should root fairly quickly and easily.

There is really no limit to the number of stems or branches that can be pegged down around each plant – except, maybe, the amount of ground available. Commercial nurserymen generally cut down young parent plants to ground level to encourage them to throw up a mass of young whippy shoots. All of these, with the exception of the weakest, are then pegged down, generally in a circle. While these are rooting a further crop of young shoots is being produced by the parent plant ready for layering when the previous crop has been lifted.

Preparing the soil Prior to layering, the soil around the parent plant must be adequately prepared to ensure quick rooting and a really substantial root system. Dig the ground to about the depth of a fork and break it down as finely as possible. There should be no hard lumps or clods as these will make layering difficult. If the ground is very sodden, as is often the case with clay soils, then work in a good quantity of coarse horticultural sand or grit. This will keep it 'open', ensuring good drainage of surplus water and adequate aeration. A stem will not root if the ground is very wet or lacks adequate oxygen.

A light sandy soil is ideal for layering as generally it is well drained, well aerated and warms quickly in the spring. However, such a soil could be so well drained that it dries out rapidly in hot dry weather. In this instance it would be advantageous to work into the soil a good quantity of moist sphagnum peat as this will hold moisture during dry weather.

If you have a soil somewhere between the two extremes of heavy wet clay and sand then I recommend that you incorporate a quantity of cutting compost into the soil; this is simply a mixture of equal parts by volume of moist sphagnum peat and coarse horticultural sand or grit.

When using any of these materials to improve the existing soil make sure they are really well mixed with the garden soil to a depth of about 30 cm (12 in.). This will ensure that the layers produce a really deep root system which is ideal for young plants which are to be transplanted. It ensures better establishment.

After preparing the layering site by digging and adding peat and/or sand, it should receive moderate firming with the heels. Do not over-firm, though, so that you make the ground rock hard as it will then be difficult to carry out layering and the stems may have difficulty in rooting into the soil.

The method Now to the actual technique of layering a stem or shoot. To encourage rooting in the fastest possible time (fast, that is, for layering), it is necessary to wound the part of the stem which is to be in contact with the soil. This results in healing tissue or callus being produced which is then followed by root production.

Make the wound about 30 cm (12 in.) from the tip of the shoot. There are various ways of making a wound in the stem. The most popular is to make a cut about 5 cm (2 in.) long halfway through the stem. When this cut is opened up a 'tongue' is formed. This cut should be kept open by inserting a small stone or piece of wood. Generally I would recommend the cut is made through a leaf joint or node as it is in this region that roots are most quickly formed by the plant.

Another way of wounding the stem is to grip it firmly in both hands and give it a sharp twist. This will have the effect of breaking some of the tissues.

The stem should then be pegged down into a shallow hole in the soil. A depression about 10–15 cm (4–6) in.) is generally adequate. Pull the stem down so that the wounded part is in close contact with the soil in the bottom of the hole and then firmly peg it down with a galvanized wire pin – a layering pin or peg is easily made from heavy gauge galvanized wire bent to the shape of a hairpin – or a lump of stone or concrete could be used to hold down the layer instead. When pegging down the stem ensure the cut remains open.

If the stem has a tendency to spring upwards and an ordinary layering pin will not hold it, I would suggest you drive in a wooden stake and tie the stem to it. Then you should be able to insert the layering pin in the normal way.

When the layer has been pegged down, insert a short cane and tie the end of the stem to it with soft garden twine or raffia to hold it in a more or less upright position. This ensures that you have a straight-stemmed young plant when the rooted layer is lifted.

The part of the stem in the hole or depression should now be covered with soil. It could with advantage have 15 cm (6 in.) of soil placed over it and lightly firmed, which will help to keep the layer moist. A shallow covering of soil will result in the layer becoming dry and of course this will inhibit rooting. (See Figure 41.)

Aftercare of the layers The layers must be kept moist at all times. Inspect them regularly during dry weather and water thoroughly with a rosed watering can or hosepipe if the surface of the soil is starting to dry out. It will also be necessary to keep weeds under control as the layers should not be smothered. It is best to hand weed around layers as chemical weedkillers could have an adverse effect on rooting. Do not use a hoe too near layers for fear of pulling them out of the ground.

The layers should be well rooted in eighteen to twenty-four months, depending on the species. At this stage the layers are severed from the parent plant and then very carefully lifted with a fork, taking care not to damage the roots. It is best to try to lift rooted layers during the autumn or early spring as these are the best times for transplanting. Try to replant immediately, but in any case do not allow the roots to dry out. If you are unable to replant straight away, the layers should be heeled in on a spare piece of ground. After planting, the young plants should be kept moist, so water them whenever the surface of the soil starts to become dry.

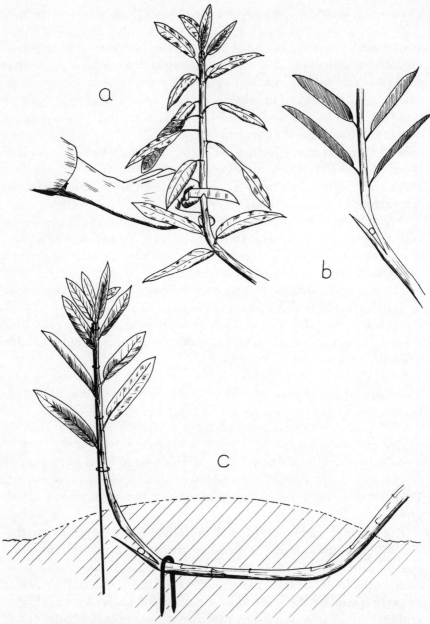

Figure 41 Simple layering of a *Rhododendron*. (a) Cut a tongue in the stem to be layered and (b) keep it open with a piece of wood or a small stone. (c) The stem is pegged down into the soil and held in place with a wire peg. A short cane keeps the top of the stem upright. The part of the stem pegged down should be covered with about 15 cm (6 in.) of soil

Many trees and shrubs can be layered but in the tables I have generally indicated layering only for subjects which may prove difficult for the amateur gardener to propagate by other methods. What I would suggest is that if you have a plant that you want to layer, but find this method is not indicated in the tables for the particular subject, try layering as there is a very good chance that it will work.

12 Other methods of layering

Simple layering, as described in Chapter 11, is by no means the only way of layering plants. There is a method whereby the stems are rooted without coming into contact with the soil, and this is known as 'air-layering'. Then there is 'serpentine layering' in which long stems are pegged down to the soil in a number of places along their length. 'Tip-layering' is where the tips of shoots are pegged down into the soil. Strawberries (*Fragaria*) are propagated by pegging down the runners and the shoots of carnations (*Dianthus*) can be pegged down to root in the soil. There are several greenhouse plants which respond to growths being pegged into the soil to root. In this chapter I will describe all of these techniques in detail.

Air-layering

Trees and shrubs This method is used for those trees and shrubs whose branches cannot be pulled down to ground level as with simple layering. It is carried out during the same period as simple layering, that is, April to August, when the plants are growing.

Old trees and shrubs can be air-layered as well as young specimens, but ensure that you use only young shoots; these root more easily than old wood. The shoot must first of all be wounded approximately 30 cm (12 in.) from its tip to encourage rapid rooting. The best method of wounding is to cut the shoot half way diagonally through its length to form a tongue about 5 cm (2 in.) in length. Dust the cut surfaces with a hormone rooting powder, preferably one formulated for semi-ripe or hardwood cuttings.

The cut must be kept open and this is easily achieved by packing it with a wad of moist sphagnum moss which you should be able to obtain from a florist. This wounded part of the shoot should now be wrapped with a bandage of moist sphagnum moss – this acts as a rooting medium and roots will rapidly develop in the moss.

The bandage of moss is held in place by wrapping it with a piece of clear polythene sheeting. Both ends of this polythene sleeve should be tightly sealed with a few twists of self-adhesive waterproof tape. The overlapping edge of the polythene sleeve should also be sealed with a strip of tape. If water gets into the sleeve the moss will become excessively wet and this may affect rooting of the shoot.

The air-layer can now be left until white roots start to show under the polythene. The length of time that shoots will take to root depends on the species, but shoots are usually well rooted within twelve months.

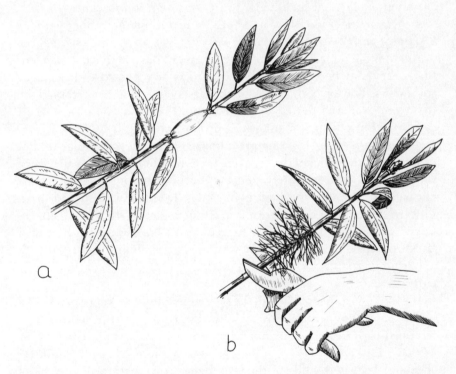

Figure 42 Air-layering a *Rhododendron*. (a) After a stem has been wounded by cutting a tongue in it, the prepared part is wrapped in sphagnum moss which is then held in place with a 'bandage' of clear polythene sheeting. (b) Generally the shoot should be well rooted within twelve months, when it can be unwrapped and severed from the parent plant

When roots are seen the polythene should be removed, together with the moss, taking care to avoid root damage. Cut the rooted shoot away from the parent plant, just beyond the rooted area. I then prefer to put the rooted layers into pots of a suitable size, and establish them in cold frames before planting. Alternatively, plant them direct in the garden or nursery bed. The most suitable time of the year to remove rooted shoots is either in the autumn or in early spring.

Air-layering is used for those subjects which are normally difficult to propagate by other methods, like cuttings or seeds. As with simple layering, you will very quickly obtain a young plant of reasonable size.

Greenhouse plants There are a number of greenhouse plants which are difficult for the amateur to propagate by other means such as cuttings, and air-layering provides an easy alternative. Plants which are often air-layered include croton (*Codiaeum*), dumb cane (*Dieffenbachia*), *Dracaena*, *Ficus* of various species, including the rubber plant or *F. elastica*, and *Philodendron*.

Another good reason for air-layering greenhouse plants is that eventually they may become too tall and need to be replaced with younger specimens. The top of the plant can be air-layered, and when rooted the top is removed and potted separately, and the remainder of the plant can be cut back to more manageable proportions.

The method of air-layering greenhouse plants is very similar to that described in the previous section on 'Trees and Shrubs'. Near the top of the plant, say approximately 30 cm (12 in.) down from the top, make a cut halfway through the stem about 7 cm (3 in.) in length and in an upward direction. This will result in a tongue being formed. An alternative way to wound the stem is to remove a 1 cm (½ in.) wide ring of bark from around the stem. The wounded area should be dusted with a hormone rooting powder formulated for semi-ripe or hardwood cuttings.

If a tongue has been made in the stem this should be kept open with a wad of moist sphagnum moss. The wounded part of the stem is then wrapped with a bandage of moist sphagnum moss and this can be kept in place by completely wrapping it with a piece of clear polythene sheeting. Seal both ends of the polythene sleeve with a few twists of waterproof self-adhesive tape and also seal the edge of the polythene with a strip of tape.

If the plants are kept in a warm humid atmosphere such as provided by a heated greenhouse, the stems should be well rooted within a

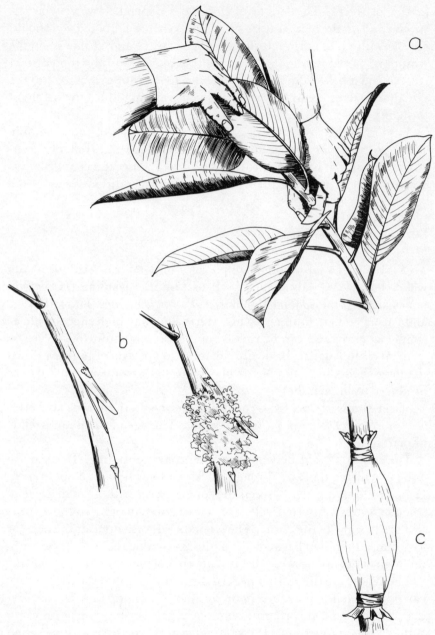

Figure 43 Air-layering a rubber plant (*Ficus elastica*): (a) A tongue is cut in the stem, about 30 cm (12 in.) from the top. (b) The tongue can be kept open with a wad of moist sphagnum moss. (c) The prepared part of the stem is then wrapped with a 'bandage' of moist moss which is held in place with a piece of clear polythene

matter of weeks. The length of time does depend, however, on the species and on the temperature provided. When well rooted you should see the white roots through the polythene sleeve and at this stage the wrapping materials should be carefully removed and the rooted stem cut away just below the roots. If you wish to save the original plant the remaining stems could be cut back if you want to reduce the height of the plant.

As soon as the rooted shoot has been removed pot it into a suitable size of pot and then keep it in a warm humid place that is shaded from strong sun until it has become well established. Once the young plant is growing well it can be placed in normal growing conditions.

Serpentine layering

This is a slight adaptation of simple layering and it is used for plants which have very long stems, such as *Clematis*, jasmine (*Jasminum*), honeysuckle (*Lonicera*), passion flower (*Passiflora*), vines (*Vitis*), *Wisteria* and various other climbing plants. It has the same high success rate as simple layering and can be carried out at the same time of the year – from April to August. Basically the method consists of pegging down the long shoots in a number of places so that several new plants are produced from each shoot.

Select young shoots for serpentine layering as these root far better and more quickly than old woody stems. The age of the plant itself is immaterial.

The shoots should be wounded to encourage rooting. This can be done by cutting halfway through the shoot for a length of about 5 cm (2 in.) to form a tongue. Keep the tongue open with a small piece of wood or a stone. Alternatively sharply twist the shoot to wound it – this will break some of the tissue. The wounding is best done at the node or leaf joint rather than between the nodes, as rooting seems to be better in this area. The shoot is wounded in a number of places along its length.

The shoots are then either pegged down directly into the soil, or they can be pegged into pots sunk in the ground. If pegging into the soil, first dig this over and incorporate some peat and coarse sand to make a suitable rooting medium. If rooting into pots, fill these with a cutting compost – a mixture of equal parts peat and coarse sand.

If rooting into the soil, peg the shoots down into a slight depression at the points where they were wounded and cover with a layer of soil.

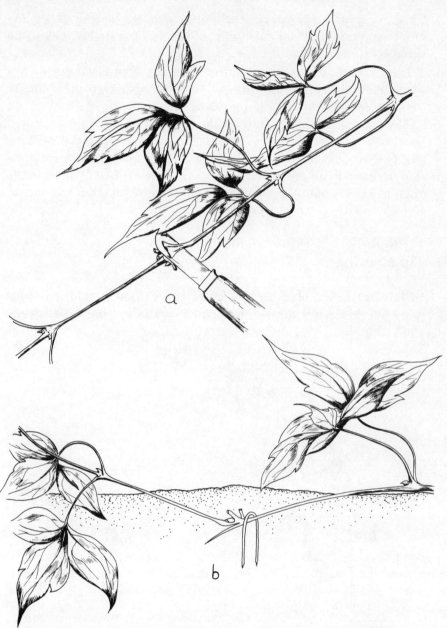

Figure 44 Serpentine layering a *Clematis*. The long shoots are prepared and pegged down in a number of places along their length so that several new plants are produced from each shoot: (a) Preparation consists of cutting a tongue through a node or leaf joint, having first stripped the leaves in that area. (b) The prepared parts of a shoot are then pegged down into the soil, using a wire peg

Galvanized wire pegs can be used in the shape of a hairpin. If pegging into pots, ensure the wounded part of the shoot is slightly below the surface of the compost.

It is important to ensure that the layers are kept moist during dry weather otherwise rooting may be delayed. Also keep them free of weeds by carefully hand weeding among them.

The rooting period depends on the subject but a good root system should have formed on most subjects within twelve months, perhaps less. Once rooted carefully lift the layers and cut away the rooted parts of each shoot. Plant the rooted portions in a nursery bed or direct in the ornamental garden. And, of course, they can be potted if you intend giving away the plants as presents.

Tip-layering

In July or August, blackberries, loganberries (*Rubus*) and the various other hybrid berries can be propagated by tip-layering. These fruits

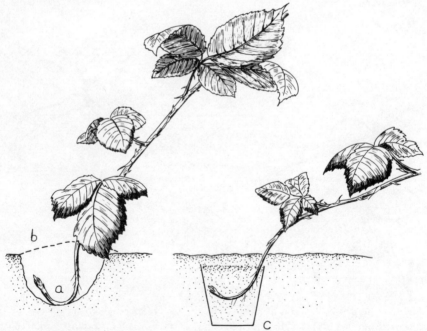

Figure 45 Tip layering a blackberry: (a) The extreme tip of a shoot is buried in the soil and (b) it should be covered with about 7.5 cm (3 in.) of soil. (c) Alternatively the tip of a shoot could be inserted in a pot of cutting compost sunk in the soil

make long 'canes' or stems and you should choose only young stems for layering– those produced in the current season.

First of all loosen the soil where the tips of the stems are to be buried, to facilitate rooting. The extreme tip of a shoot or cane is then buried in the soil to a depth of about 7.5 cm (3 in.), completely covering the tip of the stem. Generally it is not necessary to use a peg to hold down the stem. If you prefer, the tip of a stem could be inserted in a pot of cutting compost sunk up to its rim in the soil. This will minimize root disturbance when the rooted layer is lifted.

As with simple layering, the soil must be kept moist at all times to encourage quick root formation. Rooting is indeed fairly rapid, and by November the tips of the shoots should have an adequate root system to allow lifting and transplanting. Sever the rooted portion just beyond the roots and plant elsewhere in the garden. Or, if you prefer, the rooted layers could be left until the spring of the following year before being lifted. Blackberries, loganberries and other hybrid berries generally take about two years from propagation to bear fruit.

Layering shrubs by dropping

Some of the small low-growing shrubs like the heaths and heathers (*Erica*, *Calluna* and *Daboecia*), and *Gaultheria*, *Pernettya* and *Vaccinium* can be increased by a method of layering known as 'dropping'. This results in quite a large quantity of new plants, which is often desirable with the subjects mentioned as mass-planting is frequently necessary for ground cover. This form of layering is generally carried out in April and by the autumn of the same year rooted shoots can be lifted.

The plants to be increased should be carefully lifted with the minimum of root disturbance and replanted in a deeper hole so that about the lower third to half of the stems is buried. It is important to spread out the stems and work some fine soil between them. Then firm the soil well with your fingers. As with other forms of layering it is necessary to keep the plants well watered during dry weather.

In the following autumn lift the plants and cut off all the rooted stems. These rooted portions can either be potted or they can be planted direct in the garden. It is up to you to decide if the original plant is worth keeping. It should survive and grow again if replanted – I suggest that you cut back the remaining stems and then there is the possibility they will produce new growth in the spring. In any case, you will now have a good number of new plants so it may not be worth worrying about the original plant.

Strawberry runners

Strawberries (*Fragaria*) are able to propagate themselves vegetatively by means of runners, thin stems which grow over the surface of the soil and root from buds which are produced at intervals along them. Wherever part of the stem roots into the soil a new plant is produced. If you grow strawberries you will know that the plants generally produce quite a few runners and young plants develop all around the parent plants. If the runners are not required for propagation purposes they should be cut off at an early stage. It is, however, a good idea to have a succession of young strawberry plants coming along, for after about three years the plants start to decline in vigour and crop less freely, and therefore need replacing with young plants.

To speed up the process of rooting, gardeners generally peg down the runners, or layer them. The runners are produced in summer and a good time for layering them is in June and July. As this is often a dry time of the year it is of the utmost importance to keep the soil moist at all times. You will find that rooting is very poor if the soil is dry.

Before propagating new plants from runners first make sure the parent plants are absolutely free from virus diseases. It is not a good policy to propagate from virus-infected plants – indeed the affected plants should be removed and burned as there is no cure. Virus-infected plants will never crop as well as healthy plants. Initially it is best to buy virus-free plants from a specialist fruit grower and then to replace the entire stock with healthy plants every five years or so.

Layering the runners It is usual to layer the first plantlet on a runner – that is, the one nearest to the plant – for this is the strongest, and any subsequent plantlets are weaker in growth. Any runners which are produced beyond this plantlet should be removed. Any one parent plant will comfortably yield about five or six new plants; do not take any more than this as it could have a weakening effect so that fruiting is reduced in the following year. All other runners produced should be removed as soon as they appear.

The plantlets should be held in close contact with the soil to ensure the quickest possible rooting: insert a galvanized wire peg just behind

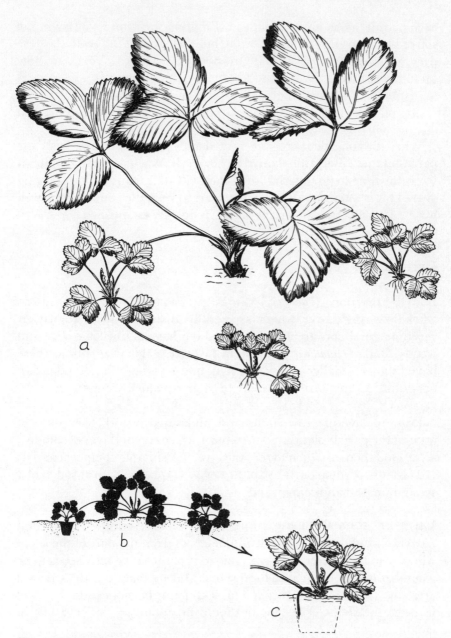

Figure 46 Layering strawberry runners: (a) Strawberries produce runners or thin stems which root into the soil and produce new plants. (b) To speed up the process of rooting, the runners can be pegged down into pots of potting compost sunk in the soil. It is usual to peg down only the first plantlet on each runner. (c) To ensure the plantlets are in close contact with the soil insert a wire peg just behind each one

each plantlet. You have the choice of rooting the runners direct in the soil or in pots. With the first method, lightly fork over the ground before pegging down the runners. With the pot method sink 7.5 cm (3 in.) pots up to their rims in the soil and fill them with potting compost or good garden soil. Then peg the runners into these. With this method the young plants will suffer far less root disturbance when being planted out and will become established very quickly.

When the plantlets are well rooted they can be severed from the parent plants, lifted and planted in new beds. August or early September is the time to do this, and then they will start fruiting the following year. This is often a dry time of the year, so keep the young plants well watered if necessary so that they become quickly established.

Carnation layering

Border carnations (*Dianthus*) can be propagated by layering shoots when flowering is over, which is generally by the middle of August. At this time the shoots are well developed but have not become hard and woody. Pinks (*Dianthus*) can also be layered at this time, but as these have thinner, smaller shoots you may find layering a bit of a delicate operation. In this instance resort to pipings, which were discussed in Chapter 7.

Prior to layering carnations and pinks lightly fork over the soil around the parent plants and then place a 5 cm (2 in.) layer of compost all around them in which to root the layers. A suitable compost consists of a mixture of equal parts by volume of good-quality loam or soil, moist sphagnum peat and coarse sand.

Layering shoots Choose strong shoots for layering – those that have not produced flowers. Strip the leaves from the part of the shoot which is to be covered with soil; this part of the shoot also needs to be wounded to encourage it to form roots. Make a tongue in the stripped area by cutting right through a leaf joint or node and halfway through the shoot. The tongue should be about 3.5 cm (1½ in.) in length.

The prepared part of the shoot should then be pressed into the layer of compost – ensuring that the tongue remains open – and secured with a galvanized wire peg. Cover the pegged part of the shoot with a 5 cm (2 in.) layer of the compost used during soil preparation.

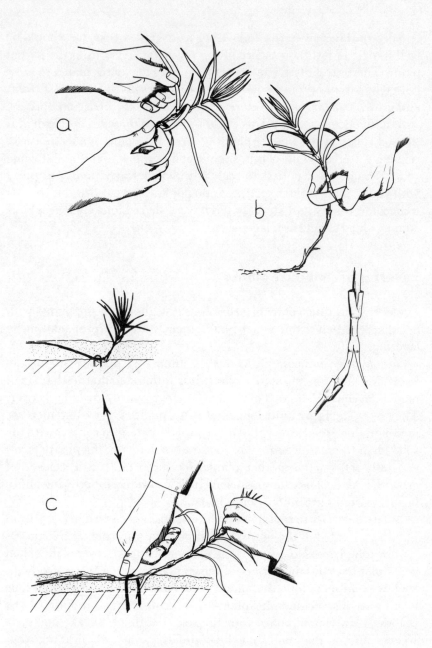

Figure 47 Layering a border carnation: (a) Strip the leaves from the part of the shoot which is to be covered with soil. (b) Cut a tongue in the stripped area, through a node or leaf joint. (c) The prepared part of the shoot is pressed into a layer of potting compost placed around the parent plant. Ensure the tongue remains open and secure with a wire peg. The pegged part of the shoot should be covered with a layer of compost

Keep the layers watered and within six to eight weeks they should be well rooted. A few days before lifting sever the layers from the parent plant. Then carefully lift them with a fork. The young plants may be potted into 9 cm (3½ in.) pots, using John Innes potting compost No. 1, and grown to a larger size before setting them in the beds and borders. Alternatively the young plants could be planted direct in the garden. If the soil is dry at the time of planting, keep the young plants moist by regular watering. The plants should come into flower the following year. Remember it is best to regularly replace carnations and pinks with young plants; older specimens do not flower so well and they start to become woody and straggly. So have a succession of young plants coming along to act as replacements.

Layering greenhouse plants

Under this heading come miscellaneous greenhouse or pot plants with peculiar characteristics which make them suited to propagation by layering.

Chlorophytum immediately comes to mind and many readers may know this better as the spider plant. It is rather peculiar in that small plants develop on the tips of the flower stems when flowering is over. The flower stems are quite long, and as the plantlets grow and increase in weight they weigh down the stems until they come in contact with the soil. Then the plantlets quickly root into it. Actually the plantlets are generally left on the mother plants, for they result in a cascade of growth, making a pendulous specimen which is ideal for growing above head level in the greenhouse or indoors.

But if you prefer to root these plantlets, it is best to do it in a 7.5–9 cm (3–3½ in.) pot. Fill the pot with potting compost, such as a loamless type or John Innes No. 1, and stand it alongside the mother plant. Then peg a plantlet onto the surface of the compost using a piece of galvanized wire formed into the shape of a hairpin. The peg should be positioned just behind the plantlet, to ensure that the base of the plantlet is in close contact with the soil. Rooting will take up to six weeks, but you may find that the plantlets are already forming roots before being pegged down.

When the young plants are well rooted in their pots they can be cut away from the mother plants. Now they are independent, and as soon as the initial pots are full of roots the plants should be put into a larger pot. Plants can be pegged down at any time of the year, although rooting will obviously be quicker in the summer.

Saxifraga stolonifera (also known as *S. sarmentosa*) is rather like a strawberry in its method of vegetative reproduction, for it produces many plantlets on the ends of long thin stems or runners. Because it produces a great many plantlets each year it has been appropriately named mother of thousands. This plant is also best grown in a high place to allow the plantlets to cascade downwards. It can be increased in the summer in the way described for *Chlorophytum*. I should wait about six weeks before severing the rooted plantlets from the parent plant – this will ensure that they are really well rooted in the compost.

Tolmiea menziesii is an even more fascinating plant, hardy in milder parts of the country, and with the popular name of pick-a-back plant because plantlets are produced at the bases of the mature leaves. This is mainly a foliage plant; it does flower, but the blooms are rather insignificant.

Figure 48 Layering a *Chlorophytum*. The small plants produced at the tips of the flower stems can be pegged into pots of compost to encourage them to form roots

Other methods of layering

These plantlets are produced in the summer and therefore this is the best time to propagate. The method I use is to peg down a leaf which has produced a plantlet into a small pot placed alongside the parent plant. The pot should be filled with loamless or John Innes potting compost. It is only a matter of weeks before the plantlet roots into the compost, at which stage it should be cut from the main plant.

Hedera, or species and cultivars of ivy, can be increased by layering the shoots in the spring – say May or June. Ivies often produce aerial roots on the stems or shoots and they use these to cling to rough surfaces for support. But if these aerial roots come in contact with soil they will take root; these portions can be pegged down into a pot. Ideally the shoots should be pegged down fairly near the tips. Even if your ivies are not developing aerial roots I should still peg down a few shoots if you want some new plants, for roots will then almost certainly be produced. Peg down a shoot at the node or leaf joint. Once you are satisfied that the shoots are well rooted in their pots they can be cut away from the mother plant.

You will find that various other greenhouse plants often start to produce roots on the stems or shoots. Some typical examples of plants that do this are *Tradescantia* species and cultivars and *Zebrina pendula*. Although these are very easy to root from cuttings, layering presents another alternative. You will notice the shoots start to produce roots at the nodes or leaf joints so it is in these areas the shoots should be pegged down into a pot of compost. In just a few weeks the shoots will have rooted and can be severed from the main plant.

Monstera deliciosa and species of *Philodendron* also produce aerial roots on the stems, and if you can manage to pull a shoot down to soil level and peg it in the region of the aerial roots, it should quickly establish in the compost. Again I would peg the shoot into a suitable sized pot placed alongside the main pot. Do not remove the shoot from the parent plant until it is well rooted in its pot. Ideally shoots should be pegged down near their tips, if there are aerial roots in this area.

13 Nursery management

This may seem rather a strange chapter title in a book which is aimed primarily at amateur gardeners who may be propagating plants in fairly small quantities. It conjures up visions of commercial nurseries with field after field of trees, shrubs, conifers, roses and so on in their thousands.

But this chapter deals with nothing so vast as this. Quite simply, I felt that some information was needed on the care of young plants *after* the propagation period. It is all very well raising new plants but how do you care for them afterwards? I hope that this chapter will enlighten readers and I am sure all will agree that it is better to give a complete picture of plant production rather than stop half-way through the procedure, for young specimens need careful and correct attention if they are to develop into mature plants. In fact many young plants are very vulnerable to things like soil and weather conditions and could soon perish if not given a suitable environment. This chapter also goes into further detail about subjects like potting, hardening, planting and so on that have been only briefly mentioned in the other chapters.

Potting This is the correct term for the initial potting of things like rooted cuttings, seedlings and so on. Many people refer to this stage of potting as 'potting up', but this is really a meaningless term when you think about it. Whether you are potting greenhouse plants or hardy plants, you should always use small pots, just large enough to take the root system without cramping it. Over-large pots are a mistake as plants do not like a large volume of compost around their roots: it will hold more water than the plants need and it is liable to remain too wet. This could inhibit rooting and may even result in rooted cuttings, seedlings and the like rotting.

But what do I mean by small pots? I think it safe to say that for the majority of plants pots in the region of 7.5 cm (3 in.) to 9 cm (3½ in.) in diameter are suitable for the initial potting.

Most pots are now plastic; the clay ones are becoming a thing of the past. Plastic pots are obtainable either as the conventional round shape or square. I am particularly in favour of square pots as they take up far less space than rounds when they are placed 'pot thick' (that is, touching each other).

A suitable compost for potting is the next consideration. It should not be too rich in fertilizer content and therefore John Innes potting compost No. 1 is recommended. Or you may prefer to use one of the modern soilless composts, in which case choose one which is roughly equivalent to JIP1 in its fertilizer content. The instructions on the bag generally indicate the suitability of the compost for various purposes. Potting composts can be purchased from most garden centres, garden shops and stores and they come in bags of various sizes. There is no doubt that it is cheaper to buy composts in large quantities, but do ensure in this case that you are able to use them all up within a few weeks of purchase, as if kept for too long they can deteriorate.

Another point you should consider when buying compost is whether it is acid or alkaline, that is, whether or not it contains lime or chalk. Most composts do contain chalk and this is fine for the majority of plants; but there are some plants which definitely need acid or lime-free soils and for these one should use only an acid compost. It is possible to buy acid John Innes composts, as well as soilless types. Examples of genera which need acid soil include *Rhododendron, Camellia, Pieris, Pernettya, Gaultheria, Erica, Calluna, Daboecia, Kalmia* and some species of *Hydrangea* and *Magnolia*. If in doubt about whether or not plants like acid or alkaline conditions refer to a good gardening encyclopaedia.

Now we come on to the technique of potting. Firstly I should mention that nowadays it is generally considered unnecessary to place a layer of drainage material in the bottom of the pots, such as the traditional layer of 'crocks' or broken clay flower pots. Most plastic pots have adequate drainage holes in the base and therefore drainage is good.

All young plants, like rooted cuttings and seedlings, should be carefully lifted from their boxes or other containers, as often the new roots are brittle. The box or pot can be tapped on the bench to loosen the compost and then the whole lot gently eased out on to the bench, when you can then readily separate the young plants. If you find that roots are exceptionally long there is no harm in cutting them back to more manageable proportions. Cutting back roots by about one-third to half their length is perfectly satisfactory. This often stimulates a good fibrous root system in many plants like shrubs, conifers and so on.

To pot a young plant, first put a layer of compost in the bottom of the pot and firm it moderately with your fingers. Place the plant so that it is central in the pot and the roots are dangling downwards and are not cramped, then fill up to the top with more compost. Tap the pot on the bench to get the compost well down, and finish by firming all round with your fingers. You may need to add a little more compost, but make sure that you leave space at the top of the pot for watering – do not take the compost right up to the rim of the pot. Having mentioned moderate firming with the fingers, I should say that some of the soilless composts do not need any firming, so be sure to refer to the manufacturer's instructions on the bag; the traditional loam-based types, like John Innes, do need moderate firming, however.

Finally I like to water newly potted plants to settle the compost around them – this is especially necessary with loamless or soilless composts. (See Figure 49.)

Potting on Potting on involves moving a plant from a smaller pot to a larger pot, generally to the next size of pot, say from a 7.5 cm (3 in.) pot to a 10 cm (4 in.) pot; or from a 12.5 cm (5 in.) pot to a 15 cm (6 in.) pot. As mentioned under potting, the aim is to avoid a very large volume of compost around the roots. The time to pot on is when the present pot is full of roots but before the plant becomes pot bound, that is, when the pot is crammed full of roots; plants in this condition do not make good growth. The best way to find out whether or not a plant needs potting on is to invert the pot, tap the rim on a bench to loosen the soil ball and slide the pot off so that the root development can be seen clearly.

Generally when potting on a stronger compost is used – in other words, one which contains more fertilizer than the original potting mixture. A good example is John Innes potting compost No. 2, or you can use an equivalent soilless type. Again make sure you use an acid compost for lime-hating plants.

If you find that the roots are very tightly packed when you remove the plants from their existing pots, then they can be gently teased out, avoiding damage as much as possible, to ensure that they quickly root into the new compost. The actual technique of potting is the same as described before. (See Figure 49.)

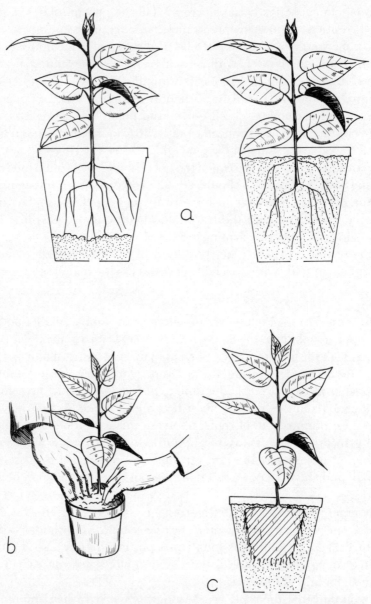

Figure 49 Potting a rooted cutting: (a) Place a layer of compost in the base of the pot
and position the plant so that the roots are dangling downwards and are not cramped.
Fill up with compost and (b) firm with the fingers, but be sure to leave space at the top
of the pot for watering. (c) A plant which has been potted on to a size larger pot. The
aim is to avoid a very large volume of compost around the roots

Hardening plants Plants which have been raised in heat and which are intended eventually to be planted out of doors must be gradually acclimatized to outdoor conditions. This process is known as hardening. The idea is to avoid giving plants a nasty shock as would happen if they were to be suddenly transferred from a high temperature to much cooler conditions.

If plants have been raised in a heated propagating case, on a heated bench under polythene, or in a mist unit, then the first stage is to transfer them to the normal greenhouse atmosphere, to allow them to become established after potting, or pricking out in the case of seedlings. Once the plants are obviously growing well the next stage is to transfer them to a cold frame to complete the acclimatization process. They should be moved to a frame three to four weeks before you wish to plant them out in the garden.

For the first few days the glass frame lights should be kept closed; then they can be opened very slightly during the day but closed again at night. Over a period of several weeks the lights can be gradually opened wider during the day until eventually they can be left off altogether, but they should still be closed at night. Then, several days before you wish to plant out, the lights can be left off at night too.

If the plants you are hardening are half-hardy annuals and perennials intended for outdoor summer bedding, then remember that these should not be planted out until all danger of frost is over, which is in late May or early June. If a frost is forecast when these subjects are hardening in cold frames, then it is wise to cover the lights at night with matting or some similar material to ensure the plants are not frosted. Do not leave the lights open at night until you are sure that there is no longer danger of frost.

Site for a nursery Even if you are raising plants on a small scale it is a good idea to have a piece of ground for the sole purpose of a nursery bed for hardy plants. Very often young trees, shrubs, conifers, climbers and roses do not look quite right in the garden until they have made larger specimens. So choose a small plot somewhere in your garden for a nursery. Ideally the soil should be deep and fertile to ensure good growth from your young plants. Above all it must be well drained, as young plants especially can suffer and even rot and die if the soil remains excessively wet over the winter. Do ensure that the soil is acid if you intend growing lime-hating plants. Often people are given cuttings or seeds of lime haters only to realize later that their soil is alkaline!

The site should be sheltered from cold winds for many plants are more tender when they are young: I can think of *Acer palmatum* and *A. japonicum*, many *Magnolia* species, *Pieris* and *Garrya elliptica*, to mention but a few.

You should also consider sun and shade. The majority of plants prefer a sunny site, but there are others that particularly like dappled or semi-shade and for these perhaps you could site your nursery so that they benefit from the shade of a tree or trees. Many commercial nurserymen, in fact, choose a site like this for growing such subjects as *Rhododendron, Camellia, Pieris, Kalmia* and other woodland shrubs.

Planting hardy plants in the nursery Throughout the book I have generally recommended potting young plants prior to setting them out in the nursery, or indeed in the garden. In many instances it is better to establish the plants in pots before setting them in the open ground. For instance if they have been raised in heat, it is more convenient to harden them in cold frames if they are in pots. But any subjects which have been raised in the open can be planted direct into nursery beds, particularly hardwood cuttings and perhaps layers.

It is really up to you to arrange your plants in the nursery, but I prefer to plant out in rows so that the plants can be easily weeded and given other types of attention. Before planting I like to thoroughly dig the soil and incorporate a good quantity of well-rotted farmyard manure or garden compost. Then I prick a general-purpose fertilizer into the surface of the soil.

Small plants are best planted with a trowel and larger plants with a spade. Always ensure that the planting hole is large enough to take the roots without cramping and make sure the roots are well down in the soil. Plants in the open ground need thorough firming to prevent wind-rock and loosening: firm them by treading all round with your heels.

If the soil is dry after planting water the plants really thoroughly, watering thereafter whenever the soil starts to become dry on the surface. It is especially important to ensure that the plants have sufficient moisture during the first spring and summer after planting. To help conserve soil moisture during dry periods you could place a 5–7.5 cm (2–3 in.) layer of organic matter around the plants in the spring. This is known as a mulch and suitable materials include well-rotted farmyard manure, garden compost, peat, leafmould, spent hops or mushroom compost (which contains lime or chalk). A mulch will also help to suppress weed growth. Indeed, you should always ensure that young plants do not become choked with weeds as growth can be severely retarded.

The ideal times for planting young hardy plants are in the dormant season for deciduous trees and shrubs (between November and March); and September/October or March/April for evergreen trees and shrubs, including conifers. I like to plant out hardy perennials in the early spring– March or April. This is also a good time for alpines.

Evergreens, including conifers, usually benefit from being sprayed with an anti-transpirant spray after planting, as this prevents their losing too much moisture through transpiration before the roots have become active. Anti-transpirant sprays are obtainable from good garden centres and shops. The alternative is to spray the plants daily with plain water for the first few weeks after planting.

Do not forget to control pests and diseases by spraying with a suitable insecticide or fungicide, for young plants are less able to withstand attacks than larger specimens. Mildew and other fungal diseases can completely cripple the few young shoots which small plants possess. Roses are particularly prone to attack by the diseases rose mildew, black spot and rose rust, so these should be controlled at the first signs of infection.

Training and trimming young plants I discussed the training of standard trees and half standards in Chapter 8, but there are some other groups of plants that also need staking and trimming or pruning in the early stages of their development.

Let us first of all deal with staking. Young climbers, while they are growing in the nursery plot, will need the support of a stout 1 m (3 ft) long bamboo cane. This should be inserted when the climbers have made not more than about 30 cm (12 in.) of growth. As the plants grow, keep them tied to the cane with raffia or soft garden twine.

I find that some shrubs benefit from being staked to prevent wind-rock. These include Leyland cypress (× *Cupressocyparis leylandii*) and also *Cupressus* species and cultivars, as they have a rather coarse root system and are easily loosened in the soil in the early years. In fact, I would suggest that *Cupressus* are pot grown in the 'nursery stage' rather than planted in the open ground, as they are rather prone to root disturbance and often do not transplant successfully.

Cytisus and *Genista* species and cultivars also benefit from being staked as again, due to a coarse root system, they are prone to wind-rock in the early years. Ideally these should also be pot grown to minimize root disturbance when they are being planted in their final positions in the garden. Other shrubs, like *Ceanothus* and *Pyracantha*, would also benefit from staking as these make rather tall growth, especially if being grown as wall-trained specimens.

Often young shrubs are not pruned or trimmed in the early stages of growth, but are allowed to grow upwards so that only a few lanky shoots or stems are produced instead of well-branched bushy growth. Examples of shrubs which benefit from pruning in their first growing season after propagation include shrubby *Cornus* or dogwoods, *Weigela*, *Deutzia*, *Diervilla*, *Spiraea*, *Buddleia*, *Viburnum*, *Philadelphus*, *Forsythia*, *Hebe*, *Potentilla*, *Cotoneaster* and *Berberis*. As a guide, those plants which naturally produce shoots from below or at ground level really benefit from pruning, for this will encourage many shoots to develop so that a bushy compact plant results. Because these plants have many young shoots, they will make far neater specimens which flower much more freely than shrubs with only a few long stems.

The pruning should be done at an early stage, say in the spring of the first growing season after propagation. The plants will be about 30 cm (12 in.) in height, although the height will depend on the particular plant. I generally cut this growth to about half its length with a pair of secateurs. If you look closely at the remaining stems you will find that there are many dormant growth buds; these will be quickly stimulated into growth if the young plants have been trimmed. In many plants there will also be growth buds below ground level and these too will produce new shoots when pruning has been done.

So do not be afraid of pruning young specimens – you will finish with far better plants. Admittedly not all young shrubs need pruning: climbers, most of the conifers and also heaths and heathers make acceptable specimens if left to grow naturally.

Lifting young plants Finally I should say a few words about lifting young plants from the nursery plot for planting in the garden. This should be done at the correct season: in other words, between November and March for all deciduous trees and shrubs; September/October or March/April for conifers and other evergreen trees and shrubs; and early spring for hardy perennials, say about March or April.

Evergreen trees and shrubs, including conifers, should be transplanted with a soil ball around their roots. This will mean cutting all

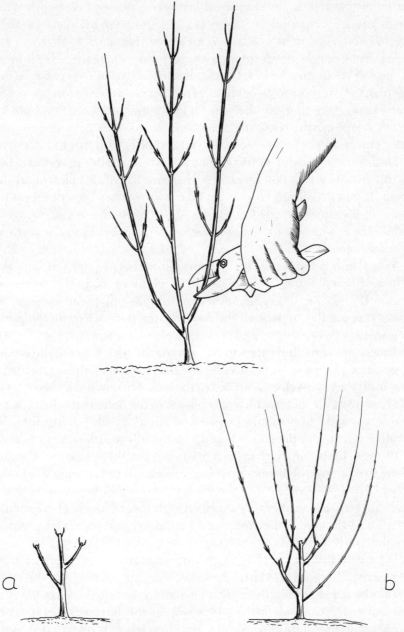

Figure 50 Pruning a young shrubby dogwood (*Cornus*) to encourage well-branched bushy growth. This can be done when the plants are about 30 cm (12 in.) in height: (a) The existing shoots can be cut back hard, cutting just above growth buds. (b) This will result in many buds breaking into growth, so producing a bushy plant

round the plant to be lifted with a sharp spade, to about the depth of the spade, and approximately 30 cm (12 in.) from the main stem. Then remove some soil to form a trench all round the plant, so that you are able to cut underneath the plant with the spade. Once the plant is loose in the soil it can then be lifted with a ball of soil around its roots. Move the plant as carefully as possible to its new site so as not to lose too much soil. If you have to move the plant a fairly long distance, it would be worth wrapping the rootball tightly with a square of hessian or something similar to contain the soil ball. Tie it tightly with twine.

Deciduous trees and shrubs can be lifted by a similar procedure, but in this instance it is not necessary to secure a ball of soil around the roots. If you are lifting large trees then you will need to cut all round them approximately 60–100 cm (2–3 ft) from the trunk to avoid damaging too much of the main root system. Then cut underneath as described above.

When transplanting on no account allow roots to dry out between lifting and replanting. If you cannot replant immediately for some reason then cover the roots with wet hessian, polythene sheeting or something similar to prevent the roots drying out. Also remember not to transplant when the ground is excessively wet or frozen.

Trees, such as standards and half standards, will generally need the support of a stout wooden stake for the first few years after transplanting, until they are well established in the soil. This should be inserted in the planting hole, before the tree is placed in the hole. It should be a 7.5 cm (3 in.) stake and should be inserted about 45 cm (18 in.) into the ground. The top of the stake should come just below the lowest branch of the tree. Once the tree has been planted it should be tied to the stake, using two or three of the modern plastic buckle-type tree-ties. Tie in the tree really tightly and ensure there is a plastic buffer between the stake and the tree (these buffers are supplied with the tree ties). There should be a tie at the top of the stake, one just above soil level and perhaps another midway between these two.

Whatever you are planting, dig a large enough hole to ensure that the roots are not cramped. With bare-rooted deciduous subjects the roots should be spread out to their fullest extent in the hole. Always plant to the same depth as when the plant was in the nursery. Never plant deeper than this or more shallowly. If a plant is on a rootstock do ensure that the graft union is above soil level, otherwise the stem of the scion will root into the soil and then the benefits of the rootstock will be lost. This is important with fruit trees especially, where a rootstock may have been chosen to control the growth of the trees. If the scion roots into the soil the rootstock will no longer have the desired growth-regulating effect.

At what stage are plants ready to be transplanted into the main garden? This is a difficult question to answer: it depends to some extent on the size of plant you wish to transplant, and some plants make sizeable specimens quicker than others. I mentioned in Chapter 8 that it takes four years to produce a standard or half-standard tree, which is then of a suitable size for transplanting. Roses are ready for transplanting in the second autumn following budding. On average, three-year-old shrubs and conifers are of a good size for transplanting. But on the other hand there are very quick-growing subjects which will make sizeable specimens in a shorter period – about two years. Herbaceous plants and other hardy perennials generally make good-sized clumps in a year from division. So use your own judgement and remember that the larger the plant – especially trees – the more difficult it will be to lift, not only due to the weight but because the roots may be growing deep in the soil.

14 A guide to the alphabetical tables

The tables have been designed so that readers will quickly be able to find a suitable method of propagating the plants in their garden, greenhouse and home.

The tables are in two sections. The first section deals with *Hardy Garden Plants* – those that are cultivated outdoors although not necessarily propagated in the open. All kinds of plants are included here, such as alpines, hardy annuals, aquatics, biennials, bulbs, corms, tubers, climbers, conifers, ferns, grasses, perennials, shrubs and trees.

The second section deals with *Greenhouse and Other Tender Plants*. Many of these are commonly grown indoors as house plants, while others are more suited to greenhouse conditions, perhaps because of their ultimate size, or because they need really good light and warm humid conditions. Also included here are frost-tender plants which are grown outdoors in the summer, like the many half-hardy annuals or bedding plants, and tender perennials, tubers and corms such as *Dahlia*, *Gladiolus*, *Pelargonium*, *Gazania*, *Canna* and the like.

The best time of year to propagate each plant is also given, together with the most suitable place or conditions. In the case of hardy plants this may be outdoors, or perhaps in a greenhouse. Where 'greenhouse' is indicated this implies that propagation generally needs to be carried out in adequate heat and humidity, such as would be provided by a propagating case, a bench heated with soil-warming cables, or maybe a mist-propagation unit. Where 'frame' is indicated, this refers to a cold frame, with no artificial heat; an alternative is the low polythene tunnel which is also recommended in the tables for some subjects.

With greenhouse and other tender plants, places to propagate again include the greenhouse. Once more, this implies that adequate heat and humidity are required, as would be provided by a propagating case, a heated bench or maybe a mist-propagation unit. Many plants can be propagated indoors, especially a large number which are grown as house plants. Where this situation is indicated it implies, ideally, a light, sunny windowsill in a warm room.

Some tender plants can be propagated in a cold frame and for these I have simply indicated 'frame'. Then you will find that a few tender plants can be propagated outdoors or in the open; for instance, dahlias can be divided just prior to planting in the spring.

The plants have been listed alphabetically under genera (their botanical names) and if a well-known common name exists this has been included also; for although botanical names should ideally be used by everyone at all times for the sake of accuracy, many people only know some of their plants by common names.

For convenience the type of plant is indicated in each instance by a key letter. The key is as follows:

A	alpine	Ca	cactus	P	perennial
An	annual	Cl	climber	S	shrub
Aq	aquatic	Cm	corm	Suc	succulent
B	bulb	F	fern	T	tree
Bi	biennial	G	grass	Tu	tuber
C	conifer	Orch	orchid		

Some species have been included separately in the tables but only when a different method of propagation from that of the genus is required.

Readers will also sometimes find the term 'cultivar' in the tables. A cultivar simply means a cultivated variety of a plant. These are plants which have been raised under cultivation and they are propagated vegetatively as opposed to seeds to ensure that they remain true to type; seedlings often have different characteristics from their parents.

Generally where the method of propagation indicates 'seeds' this is intended only for seeds from true species, which you may have collected from your plants, and not seeds from hybrid plants or cultivars. However, seedsmen often offer packets of mixed hybrids or cultivars (particularly of hardy and half-hardy annuals and perennials) so in this instance seeds provide a valid method of propagation. But do not save seeds from the resulting plants as they will not produce identical offspring.

× before a name (for example, × *Cupressocyparis*) denotes the plant is a natural hybrid (the result of cross fertilization) between species of two different genera. Such a plant is correctly called a bi-generic hybrid and must be propagated vegetatively as it will not breed true to type from seeds.

If × occurs between a generic name and a second name this also indicates a hybrid plant: in this instance the plant resulted due to cross fertilization between two or more species of the same genus. These hybrid plants should also be propagated vegetatively as they will not come true to type from seeds.

A + before a name, for example + *Laburnocytisus*, denotes the plant is a graft hybrid or chimera between species of two different genera. In this example the parent plants were *Laburnum anagyroides* and *Cytisus purpureus*. A graft hybrid or chimera is a plant derived from both the rootstock and the scion which have grown up together. The rootstock tissue has become mingled with that of the scion. With our example (+ *Laburnocytisus*) the tree produces, at random, *Laburnum* growth and flowers, *Cytisus* growth and flowers, and intermediate growth with *Laburnum*-like flowers but which are purple like those of the *Cytisus*.

With many of the plants in the tables there are various methods of propagation so choose one which suits your particular purpose and facilities available. I am not implying that one method is any better than another.

For full details of the methods of propagation, and conditions in which to carry out propagation, readers should refer to the appropriate chapters in this book.

Tables

Propagation of hardy garden plants 180–267

Propagation of greenhouse and other tender plants 268–304

Plant	Type	Method	Time	Place
ABELIA	S	Semi-ripe cuttings	July	Greenhouse
ABELIOPHYLLUM	S	Semi-ripe cuttings	July	Greenhouse
ABIES Silver Fir	C	Seeds	March	Outdoors
cultivars		Grafting (veneer)	March–April	Greenhouse
ABUTILON	S	Seeds	March–April	Greenhouse
		Semi-ripe cuttings	July–Aug	Greenhouse
ACAENA	A	Seeds	Feb–March	Frame
		Division	March–April	Outdoors/frame
ACANTHOLIMON	P	Seeds	March–April	Greenhouse
		Division	March–April	Frame
		Semi-ripe cuttings	Aug	Frame
ACANTHOPANAX	T S	Seeds	March–April	Outdoors
		Root cuttings	Dec–Jan	Frame
ACANTHUS Bear's Breeches	P	Seeds	March–April	Greenhouse
		Division	March–April	Outdoors
ACER Maple	T	Seeds	March–April	Outdoors
cultivars		Grafting (whip-and-tongue)	Feb–March	Outdoors
japonicum Japanese Maple cultivars		Grafting (spliced side veneer)	March	Greenhouse
		Softwood cuttings	April–June	Greenhouse
palmatum Japanese Maple cultivars		Grafting (spliced side veneer)	March	Greenhouse
		Softwood cuttings	April–June	Greenhouse

ACHILLEA	Yarrow	P A	Soft basal cuttings	April–May	Greenhouse
			Division	March–April	Outdoors/frame
			Seeds	May–June	Outdoors/frame
ACONITUM	Monk's Hood	P	Division	March–April	Outdoors
			Seeds	September	Frame
ACORUS	Sweet Flag	Aq	Division	April–June	Outdoors
ACTAEA		P	Division	March–April	Outdoors
			Seeds	March–April	Frame/greenhouse
ACTINIDIA		Cl	Seeds	March–April	Greenhouse
			Semi-ripe cuttings	July–Aug	Greenhouse
			Hardwood cuttings	Nov–Dec	Greenhouse
ADENOPHORA		P	Seeds	Sept or April	Frame
ADIANTUM	Maidenhair Fern	F	Division	March–April	Outdoors
			Spores	When ripe	Greenhouse
ADONIS		An	Seeds	April or Sept	Outdoors/greenhouse
		P	Division	March–April	Outdoors
			Seeds	When ripe	Outdoors
AEGOPODIUM		P	Division	March–April	Outdoors
			Root cuttings	Dec–Jan	Frame
AESCULUS	Horse Chestnut	T S	Seeds	Sept–Oct	Outdoors
			Grafting (whip-and-tongue)	Feb–March	Outdoors
			Budding	June–Sept	Outdoors

181

Plant		Type	Method	Time	Place
AETHIONEMA		A	Seeds	Jan–Feb	Outdoors
			Softwood cuttings	June–July	Greenhouse
AGAPANTHUS	African Lily	P	Division	March–April	Outdoors
			Seeds	March–April	Greenhouse
AGASTACHE		P	Seeds	May–June	Outdoors
			Division	March–April	Outdoors
AGRIMONIA	Agrimony	P	Division	March–April	Outdoors
AGROSTEMMA	Corn Cockle	An	Seeds	March–April	Outdoors
				Sept–Oct	Outdoors
AILANTHUS	Tree of Heaven	T	Seeds	March–April	Outdoors
			Root cuttings	Dec–Jan	Frame
AJUGA	Bugle	P	Division	March–April	Outdoors
AKEBIA		Cl	Seeds	Sept	Greenhouse
			Simple layering	April–Aug	Outdoors
			Semi-ripe cuttings	Aug–Sept	Greenhouse
ALCHEMILLA	Lady's Mantle	P	Division	March–April	Outdoors
			Seeds	May–June	Outdoors
ALISMA		Aq	Division	April–June	Outdoors
ALLIUM	Ornamental Onions	B	Seeds	Jan–Feb	Outdoors
			Bulblets	Aug–Sept	Outdoors
some species			Bulbils	April	Outdoors

		T S			
ALNUS cultivars	Alder	T S	Seeds Simple layering Grafting (spliced side veneer)	March–April April–Aug March	Outdoors Outdoors Greenhouse
ALOPECURUS	Foxtail	G (P)	Seeds	March	Greenhouse
ALSTROEMERIA	Peruvian Lily	P	Seeds Division (dislikes disturbance)	April April	Greenhouse Outdoors
ALTHAEA	Hollyhock	An P	Seeds	May–June Feb–March	Outdoors Greenhouse
ALYSSUM	Madwort	An P A	Seeds Seeds Semi-ripe cuttings	March March–April July–Aug	Greenhouse Outdoors Frame
AMARYLLIS		B	Bulblets	June–July	Outdoors
AMELANCHIER		S T	Seeds Simple layering Division	Sept–Oct April–May Nov–March	Outdoors/frame Outdoors Outdoors
AMORPHA	Lead Plant	S	Division Softwood cuttings	Nov–March April–June	Outdoors Greenhouse
AMPELOPSIS		Cl	Eye or hardwood cuttings Softwood cuttings	Dec June	Greenhouse Greenhouse
AMSONIA		P	Division Seeds	March–April May–June	Outdoors Outdoors

Plant	Type	Method	Time	Place
ANACYCLUS	An P	Seeds	March–April	Greenhouse
		Soft basal cuttings	April–June	Greenhouse
ANAGALLIS	P An	Seeds	March–April	Outdoors/greenhouse
ANAPHALIS Pearly Everlasting	P	Division	March–April	Outdoors
ANCHUSA Alkanet	P	Root cuttings	Dec–Jan	Frame
		Seeds	May–June	Outdoors
	An	Seeds	March–April	Outdoors
ANDROMEDA Bog Rosemary	S	Division	Nov–March	Outdoors
		Seeds	March–April	Greenhouse
		Semi-ripe cuttings	July–Aug	Greenhouse
ANDROSACE Rock Jessamine	A	Offsets/division	March–April	Outdoors
		Seeds	When ripe	Outdoors
ANEMONE Windflower	P	Division	March–April	Outdoors
		Seeds	Sept–Oct	Outdoors
x hybrida cultivars Japanese Anemone		Root cuttings	Dec–Jan	Frame
ANEMONOPSIS	P	Seeds	May–June	Frame
		Division	March–April	Outdoors
ANGELICA Angelica	P	Seeds	Aug–Sept	Outdoors
ANTENNARIA Mountain Everlasting	A	Division	March–April	Outdoors
		Seeds	Jan–Feb	Outdoors

ANTHEMIS	Chamomile	A P	Division	March–April	Outdoors
			Seeds	May–June	Outdoors
			Semi-ripe cuttings	July–August	Frame
ANTHERICUM	St Bernard's Lily	P	Division	March–April	Outdoors
			Seeds	Sept	Frame
ANTHRISCUS	Chervil	An	Seeds	March–April	Outdoors
ANTHYLLIS		S P	Semi-ripe cuttings	July–Aug	Greenhouse
			Seeds	March–April	Greenhouse
			Division	March–April	Outdoors
APONOGETON	Water Hawthorn	Aq	Division	April–June	Outdoors
AQUILEGIA	Columbine	P	Seeds	May–June	Outdoors
		A	Seeds	Jan–Feb	Outdoors
ARABIS	Rock Cress	P	Seeds	May–June	Outdoors
			Division	March–April	Outdoors
			Semi-ripe cuttings	June–Aug	Frame
ARALIA		T S	Root cuttings	Dec–Jan	Frame
			Seeds	March–April	Outdoors
ARAUCARIA araucana	Monkey Puzzle	C	Seeds	March–April	Greenhouse
ARBUTUS	Strawberry Tree	T	Seeds	March–April	Greenhouse
			Semi-ripe cuttings	Aug–Oct	Greenhouse
ARCTOSTAPHYLOS		S	Seeds	March–April	Greenhouse
			Semi-ripe cuttings	Aug–Oct	Greenhouse

185

Plant		Type	Method	Time	Place
ARENARIA	Sandwort	A	Seeds	Jan–Feb	Outdoors
			Division	March–April	Outdoors
ARGEMONE	Prickly Poppy	An P	Seeds	April	Outdoors
ARISARUM		P	Division	March–April	Outdoors
ARISTOLOCHIA		Cl	Seeds	March–April	Greenhouse
			Semi-ripe cuttings	July–Aug	Greenhouse
ARMERIA	Thrift	A	Division	March–April	Outdoors
			Semi-ripe cuttings	July–Aug	Greenhouse/frame
			Seeds	Jan–Feb	Outdoors
ARNEBIA	Prophet Flower	P	Seeds	April	Greenhouse
			Root cuttings	Dec–Jan	Greenhouse
ARONIA	Chokeberry	S	Seeds	March–April	Outdoors
			Semi-ripe cuttings	July–Oct	Greenhouse
			Division	Nov–March	Outdoors
ARTEMISIA	Wormwood	S	Semi-ripe cuttings	July–Aug	Greenhouse
		P	Division	March–April	Outdoors
ARUM		Tu	Division	March–April	Outdoors
ARUNCUS	Goat's Beard	P	Division	March–April	Outdoors
ARUNDINARIA	Bamboo	G (P)	Division	April–May	Outdoors
ARUNDO		G (P)	Division	March–April	Outdoors
ASARUM		P	Division	March–April	Outdoors

ASCLEPIAS	Milkweed	P	Division	March–April	Outdoors
			Seeds	April	Greenhouse
ASPERULA	Woodruff	P	Division	March–April	Outdoors
		An	Seeds	March–April	Outdoors
ASPHODELINE	Asphodel	P	Division	March–April	Outdoors
ASPLENIUM	Spleenwort	F	Division	April	Outdoors
			Spores	When ripe	Greenhouse
ASTER	including Michaelmas Daisy	P	Division	March–April	Outdoors
ASTILBE	False Goat's Beard	P	Division	March–April	Outdoors
ASTRAGALUS		P A	Seeds	Jan–Feb	Outdoors
ASTRANTIA	Masterwort	P	Division	March–April	Outdoors
			Seeds	May–June	Outdoors
ATHROTAXIS		C	Seeds	March–April	Outdoors
			Semi-ripe cuttings	Oct	Greenhouse/frame
ATHYRIUM		F	Division	April	Outdoors
			Spores	When ripe	Greenhouse
ATRIPLEX		S	Semi-ripe cuttings	Sept–Oct	Frame/low polythene tunnel
		An	Seeds	March–April	Outdoors
AUBRIETA		A	Softwood cuttings	April–June	Greenhouse/frame
			Hardwood cuttings	Oct–Nov	Frame
			Seeds	Jan–Feb	Outdoors

Plant		Type	Method	Time	Place
AUCUBA	Spotted Laurel	S	Semi-ripe cuttings	Sept–Oct	Frame/low polythene tunnel
AVENA	Oat	G (An and Bi)	Seeds	March–April	Outdoors
AZARA		S	Semi-ripe cuttings	Aug–Sept	Greenhouse
AZOLLA		Aq	Division	April–June	Outdoors
BALLOTA		P	Seeds	March–April	Greenhouse
		S	Softwood cuttings	April–June	Greenhouse/frame
BAPTISIA		P	Division	March–April	Outdoors
			Seeds	April–May	Outdoors/frame
BARBAREA		P	Seeds	May–June	Outdoors
			Division	March–April	Outdoors
BELLIS	Daisy	P	Division	March–April	Outdoors
			Seeds	May–June	Outdoors
double-flowered		P(Bi)	Seeds	May–June	Outdoors
BERBERIDOPSIS		S	Seeds	March–April	Greenhouse
			Softwood cuttings	April–June	Greenhouse
			Simple layering	April–May	Outdoors
BERBERIS	Barberry	S	Semi-ripe cuttings	Sept–Oct	Greenhouse/frame/low polythene tunnel

	Common name		Method (Mallet cuttings)	Sept–Oct	Greenhouse/frame/low polythene tunnel
			Seeds	March–April	Outdoors
BERGENIA		P	Division	March–April	Outdoors
			Diced rhizomes	Jan	Greenhouse
			Seeds	March–April	Greenhouse
BESCHORNERIA		P	Offsets	April	Outdoors
			Seeds	April	Greenhouse
BETULA	Birch	T	Seeds	March–April	Outdoors
			Grafting (spliced side veneer)	March	Greenhouse
BLECHNUM		F	Division	April	Outdoors
			Spores	When ripe	Greenhouse
BLETILLA		Orch	Division	After flowering	Outdoors
BOLAX		A	Division	March–April	Outdoors
BORAGO	Borage	An	Seeds	March–April	Outdoors
		P	Division	March–April	Outdoors
BOYKINIA		P	Seeds	May–June	Outdoors
			Division	March–April	Outdoors
BRACHYCOME	Swan River Daisy	An	Seeds	March	Greenhouse
		P	Division	March–April	Outdoors

Plant		Type	Method	Time	Place
BRIZA	Quaking Grass	G (An and P)	Seeds	March–April	Outdoors
				Sept–Oct	Outdoors
BRODIAEA		Cm	Cormlets	Sept	Outdoors/frame
			Seeds	Jan–Feb	Outdoors/frame
BROMUS	Brome Grass	G (An)	Seeds	March–April	Outdoors
				Sept–Oct	Outdoors
		G (P)	Division	March–April	Outdoors
BRUCKENTHALIA		S	Division	March–April	Outdoors
			Semi-ripe cuttings	June–Aug	Greenhouse/frame
			Seeds	March–April	Greenhouse
BRUNNERA		P	Seeds	May–June	Outdoors/frame
			Division	March–April	Outdoors
BUDDLEIA		S	Softwood cuttings	Jan–June	Greenhouse
			Hardwood cuttings	Nov–Dec	Frame
			Seeds	March–April	Outdoors
BULBINELLA		P	Division	March–April	Outdoors
			Seeds	Jan–Feb	Outdoors
BULBOCODIUM		Cm	Cormlets	Sept or spring	Outdoors
			Seeds	Jan–Feb	Outdoors
BUPHTHALMUM		P	Division	March–April	Outdoors
			Seeds	May–June	Outdoors

BUPLEURUM	Hare's Ear	S	Semi-ripe cuttings	June–Aug	Greenhouse
		P	Division	March–April	Outdoors
BUTOMUS	Flowering Rush	Aq	Division	April–June	Outdoors
			Seeds	April–May	Outdoors
BUXUS	Box	S	Semi-ripe cuttings	Sept–Oct	Greenhouse/frame/low polythene tunnel
			Seeds	March–April	Outdoors
CALAMINTHA	Calamint	P	Seeds	Feb–March	Outdoors
			Division	March–April	Outdoors
CALCEOLARIA	Slipperwort	A	Seeds	Jan–Feb	Outdoors/frame
			Division	April	Outdoors
CALENDULA	Marigold	An	Seeds	March–April	Outdoors
				Sept–Oct	Outdoors
CALLA		Aq	Division	April–June	Outdoors
CALLICARPA		S	Seeds	March–April	Outdoors
			Semi-ripe cuttings	June–Aug	Greenhouse
CALLIRRHOE	Poppy Mallow	P	Softwood cuttings	May–June	Frame
			Seeds	April–May	Frame
CALLITRICHE	Water Starwort	Aq	Soft cuttings	April–June	Outdoors
CALLUNA	Heather, Ling	S	Semi-ripe cuttings	June–Sept	Greenhouse/frame
			Layering (dropping)	April	Outdoors
			Seeds	March–April	Greenhouse

Plant		Type	Method	Time	Place
CALOCEDRUS	Incense Cedar	C	Semi-ripe cuttings	Sept–Oct	Greenhouse/frame
			Seeds	March–April	Outdoors
CALOCHORTUS		B	Seeds	Feb–March	Frame
			Bulbils	April	Frame
			Bulblets	Aug–Sept	Frame/outdoors
CALTHA	Marsh Marigold	Aq	Division	April–June	Outdoors
			Seeds	April–May	Outdoors
CALYCANTHUS		S	Simple layering	April–Aug	Outdoors
			Seeds	March–April	Greenhouse/outdoors
CAMASSIA	Quamash	B	Bulblets	Sept	Outdoors
			Seeds	Sept or Feb–March	Outdoors
CAMELLIA		S	Semi-ripe cuttings	Aug	Greenhouse
			Leaf-bud cuttings	Aug	Greenhouse
			Grafting (spliced side veneer)	March	Greenhouse
			Simple layering	April–Aug	Outdoors
			Seeds	March–April	Greenhouse
CAMPANULA	Bellflower	P A	Seeds	Jan–Feb	Outdoors
				May–June	Outdoors
			Division	March–April	Outdoors
			Soft basal cuttings	May–July	Greenhouse/frame
medium	Canterbury Bell	Bi	Seeds	May–June	Outdoors

CAMPSIS		Cl	Root cuttings	Dec–Jan	Frame/greenhouse
			Seeds	March–April	Greenhouse
			Semi-ripe cuttings	July–Aug	Greenhouse
CARAGANA		S T	Seeds	March–April	Outdoors
			Semi-ripe cuttings	July–Aug	Greenhouse
			Simple layering	April–Aug	Outdoors
CARDAMINE		P	Division	March–April	Outdoors
			Seeds	May–June	Outdoors
CARDIOCRINUM	Giant Lily	B	Seeds	Sept or Jan–Feb	Outdoors
			Bulblets	Sept	Outdoors
CARDUNCELLUS		P	Seeds	May	Outdoors
			Division	March–April	Outdoors
			Root cuttings	Dec–Jan	Frame/greenhouse
CAREX	Sedge	P	Division	March–April	Outdoors
CARLINA		P	Seeds	May–June	Outdoors
CARPENTERIA		S	Seeds	March–April	Greenhouse
			Simple layering	April–Aug	Outdoors
			Semi-ripe cuttings	July–Aug	Greenhouse
CARPINUS cultivars	Hornbeam	T	Seeds	March–April	Outdoors
			Grafting (spliced side veneer)	March	Greenhouse

Plant		Type	Method	Time	Place
CARYA	Hickory	T	Seeds	When ripe	Frame (in pots)
CARYOPTERIS		S	Semi-ripe cuttings	June–Aug	Greenhouse
			Softwood cuttings	April–June	Greenhouse
CASSINIA		S	Semi-ripe cuttings	Sept–Oct	Frame/low polythene tunnel
CASSIOPE		S	Semi-ripe cuttings	Aug	Greenhouse
			Layering (dropping)	April	Outdoors
			Simple layering	April–Aug	Outdoors
			Seeds	March–April	Greenhouse
CASTANEA	Sweet Chestnut	T	Seeds	Sept–Oct	Outdoors
			Grafting (whip-and-tongue)	March	Outdoors
CATALPA	Indian Bean Tree	T	Seeds	March–April	Greenhouse/frame
			Semi-ripe cuttings	Aug	Greenhouse
			Root cuttings	Dec–Jan	Frame
CATANANCHE		P	Seeds	Feb–March	Greenhouse
				May–June	Outdoors
			Root cuttings	Dec–Jan	Frame
CEANOTHUS		S	Semi-ripe cuttings	Aug–Sept	Greenhouse
			Root cuttings	Nov–Dec	Greenhouse
			Seeds	March–April	Greenhouse
CEDRUS	Cedar	C	Seeds	March–April	Outdoors
			Grafting (veneer)	March	Greenhouse

		Cl			
CELASTRUS			Seeds	March–April	Outdoors
			Simple layering	April-Aug	Outdoors
			Hardwood cuttings	Nov–Dec	Greenhouse
CELMISIA		S P	Seeds	Jan–Feb	Outdoors (in pots)
			Soft basal cuttings	April–June	Greenhouse
			Division	March–April	Outdoors
CENTAUREA	Cornflower	An	Seeds	March–April	Outdoors
		P	Division	March–April	Outdoors
		S	Semi-ripe cuttings	Aug	Greenhouse
CENTRANTHUS	Valerian	P	Seeds	May–June	Outdoors
			Softwood cuttings	April–June	Greenhouse
CEPHALARIA	Giant Scabious	P	Division	March–April	Outdoors
			Seeds	May–June	Outdoors
			Root cuttings	Dec–Jan	Frame
CEPHALOTAXUS	Plum Yew	C	Seeds	March–April	Outdoors
			Semi-ripe cuttings	Sept	Greenhouse/frame
CERASTIUM	Snow in Summer	A	Seeds	Jan–Feb	Outdoors
			Division	March–April	Outdoors
			Soft/semi-ripe cuttings	April–Aug	Greenhouse/frame
CERATOSTIGMA	Leadwort	S	Semi-ripe cuttings	Aug	Greenhouse/frame
			Division	March–April	Outdoors
		P	Division	March–April	Outdoors
CERCIDIPHYLLUM		T	Seeds	March–April	Outdoors
			Simple layering	April–Aug	Outdoors

Plant		Type	Method	Time	Place
CERCIS	Judas Tree	T	Seeds	March–April	Greenhouse
			Grafting (whip-and-tongue)	March	Outdoors
CHAENOMELES	Ornamental Quince	S	Simple layering	April–Aug	Outdoors
			Seeds	March–April	Outdoors
CHAMAECYPARIS	False Cypress	C	Seeds	March–April	Outdoors
difficult and slow-growing cultivars			Semi-ripe cuttings	Sept–Oct	Greenhouse/frame
			Grafting (veneer)	March	Greenhouse
CHAMAEDAPHNE		S	Simple layering	April–Aug	Outdoors
			Semi-ripe cuttings	July–Sept	Greenhouse
			Seeds	March–April	Greenhouse
CHAMAEMELUM	Chamomile	P	Division	March–April	Outdoors
			Softwood cuttings	April–June	Greenhouse
			Seeds	May–June	Outdoors
CHEIRANTHUS	Wallflower	P(Bi)	Seeds	May–June	Outdoors
			Softwood cuttings	April–June	Greenhouse
CHELONE	Turtlehead	P	Division	March–April	Outdoors
			Seeds	May–June	Outdoors
			Softwood cuttings	April–June	Frame
CHIASTOPHYLLUM		A	Division	March–April	Outdoors
			Softwood cuttings	April–June	Greenhouse/frame

CHIMONANTHUS	Winter Sweet	S	Simple layering Seeds	April–Aug March–April	Outdoors Greenhouse/frame
CHIONANTHUS		S	Simple layering Grafting (whip-and-tongue on *Fraxinus*) Seeds	April–Aug March March–April	Outdoors Outdoors Greenhouse/frame
CHIONODOXA		B	Bulblets Seeds	Sept Autumn or Jan–Feb	Outdoors Outdoors
CHOISYA	Mexican Orange Blossom	S	Semi-ripe cuttings	Aug–Oct	Frame/low polythene tunnel/greenhouse
CHRYSANTHEMUM	Annual Chrysanthemum	An	Seeds	March–April	Outdoors
	Early-flowering Border Chrysanthemum	P	Division Soft basal cuttings	March–April Jan–April	Outdoors Greenhouse
	Ox-eye Daisy, Shasta Daisy and similar	P A	Division	March–April	Outdoors
CHRYSOGONUM		P	Division	March–April	Outdoors
CIMICIFUGA	Bugbane	P	Division Seeds	March–April When ripe	Outdoors Frame
CIRCIUM		P	Division Seeds	March–April May–June	Outdoors Outdoors

Plant	Type	Method	Time	Place
CISTUS	S	Semi-ripe cuttings	Aug–Oct	Frame/low polythene tunnel/greenhouse
		Seeds	March–April	Greenhouse/frame
CLADANTHUS	An	Seeds	March–April	Outdoors
CLADRASTIS	T	Seeds	March–April	Outdoors/frame
		Root cuttings	Dec–Jan	Frame
CLARKIA	An	Seeds	March–April	Outdoors
			Sept–Oct	Outdoors
CLEMATIS	Cl	Leaf-bud cuttings	April–June	Greenhouse
		Serpentine layering	April–Aug	Outdoors
	P	Division	March–April	Outdoors
	Cl P	Seeds	March–April	Outdoors
CLERODENDRON	S	Semi-ripe cuttings	July–Aug	Greenhouse
		Suckers	March–April	Outdoors
		Root cuttings	Dec–Jan	Frame
CLETHRA	S T	Seeds	March–April	Greenhouse
		Semi-ripe cuttings	Aug–Sept	Greenhouse
		Division	March	Outdoors
CODONOPSIS	P	Division	March–April	Outdoors
		Seeds	May–June	Outdoors
		Soft basal cuttings	April–June	Frame/greenhouse

		G (An)	Seeds	Feb–March May	Greenhouse Outdoors
COIX	Job's Tears		Seeds		
COLCHICUM		Cm	Seeds	When ripe	Outdoors/frame
			Cormlets	July	Outdoors
COLLETIA		S	Semi-ripe cuttings	July–Sept	Greenhouse
			Seeds	March–April	Greenhouse/frame
COLLINSIA		An	Seeds	March–April	Outdoors
				Sept–Oct	Outdoors
COLUTEA	Bladder Senna	S	Semi-ripe cuttings	Aug–Sept	Greenhouse
			Seeds	March–April	Greenhouse
CONVALLARIA	Lily of the Valley	P	Division	After flowering	Outdoors
			Seeds	When ripe	Frame
CONVOLVULUS		An	Seeds	March–April	Outdoors
		S	Semi-ripe cuttings	Aug–Sept	Greenhouse
			Seeds	March–April	Greenhouse
			Root cuttings	Dec–Jan	Greenhouse
COPROSMA		S T	Semi-ripe cuttings	July–Aug	Greenhouse
COREOPSIS	Tick Seed	P	Division	March–April	Outdoors
			Seeds	May–June	Outdoors
		An	Seeds	March–April	Outdoors
CORIARIA		S	Seeds	When ripe or March–April	Greenhouse
			Root cuttings	Dec–Jan	Frame
			Semi-ripe cuttings	Aug–Sept	Greenhouse

199

Plant		Type	Method	Time	Place
CORNUS	Cornel, Dogwood	S T	Hardwood cuttings	Nov–Dec	Frame
			Seeds	March–April	Frame/greenhouse
			Suckers	March–April	Outdoors
			Layering	April–Aug	Outdoors
canadensis	Creeping Dogwood	P	Division	March–April	Outdoors
(also known as					
Chamaepericlymenum					
canadense)					
COROKIA		S	Seeds	March–April	Greenhouse/frame
			Semi-ripe cuttings	Aug–Sept	Greenhouse
CORONILLA		S	Semi-ripe cuttings	Aug–Sept	Greenhouse
			Seeds	March–April	Greenhouse
CORTADERIA	Pampas Grass	G (P)	Division	March–April	Outdoors
CORYDALIS		P A	Division	March–April	Outdoors
			Seeds	When ripe	Frame/outdoors
CORYLOPSIS		S	Simple layering	April–Aug	Outdoors
			Softwood cuttings	April–June	Greenhouse
			Seeds	March–April	Greenhouse/frame
CORYLUS	Hazel, Filbert	S T	Simple layering	April–Aug	Outdoors
			Seeds	Autumn or	Outdoors
				March–April	
COTINUS		S	Semi-ripe cuttings	Aug–Sept	Greenhouse
			Simple layering	April–Aug	Outdoors

200

Genus	Common name	Type	Method	Time	Location
COTONEASTER		S	Seeds	March–April	Outdoors
			Semi-ripe cuttings	Aug–Sept	Greenhouse
			Simple layering	April–Aug	Outdoors
COTULA		A	Division	March–April	Outdoors
			Seeds	Jan–Feb	Outdoors
CRAMBE		P	Root cuttings	Dec–Jan	Frame
			Division	March–April	Outdoors
CRASSULA		A (Suc)	Division	March–April	Outdoors
			Cuttings	Summer	Greenhouse
+ CRATAEGOMESPILUS		T	Simple or air layering	April–Aug	Outdoors
CRATAEGUS	Thorn	T S	Seeds	March–April	Outdoors
			Grafting (whip-and-tongue)	March	Outdoors
			Budding	June–Aug	Outdoors
			Simple or air layering	April–Aug	Outdoors
CREPIS	Hawkweed	P	Division	March–April	Outdoors
			Seeds	May–June	Outdoors
CRINODENDRON		T	Semi-ripe cuttings	Aug	Greenhouse
CRINUM		B	Division	April–May	Outdoors
			Bulblets	April–May	Outdoors
			Seeds	When ripe	Greenhouse

Plant		Type	Method	Time	Place
CROCOSMIA *(Montbretia)*		Cm	Division of clumps Seeds	March–April When ripe	Outdoors Frame/greenhouse
CROCUS		Cm	Cormlets Seeds	Spring When ripe or Jan–Feb	Outdoors Outdoors
CRYPTOGRAMMA	Mountain Parsley Fern	F	Division Spores	March–April When ripe	Outdoors Greenhouse
CRYPTOMERIA dwarf cultivars		C	Semi-ripe cuttings Seeds Grafting (veneer)	Sept–Oct March–April March	Greenhouse/frame Outdoors Greenhouse
CUNNINGHAMIA		C	Seeds Semi-ripe cuttings	March–April Sept	Outdoors Greenhouse
x CUPRESSOCYPARIS		C	Semi-ripe cuttings	Sept–Oct	Greenhouse/frame
CUPRESSUS cultivars difficult from cuttings	Cypress	C	Semi-ripe cuttings Seeds Grafting (veneer)	Sept–Oct March–April March	Greenhouse/frame Outdoors Greenhouse
CURTONUS		Cm	Division of clumps	April	Outdoors
CYANANTHUS		A	Seeds Softwood cuttings	Jan–Feb April–June	Outdoors Greenhouse
CYCLAMEN		Tu	Seeds	When ripe or Jan–Feb	Outdoors
CYDONIA	Quince	T	Simple or air layering	April–Aug	Outdoors

Plant	Common name	Type	Method	Time	Location
			Seeds	March–April	Outdoors
			Grafting (whip-and-tongue)	March	Outdoors
CYMBALARIA		A	Division	March–April	Outdoors
			Softwood cuttings	April–June	Greenhouse
			Seeds	Jan–Feb	Outdoors/frame
CYNOGLOSSUM	Hound's Tongue	P	Seeds	May–June	Outdoors
			Division	March–April	Outdoors
		Bi	Seeds	May–June	Outdoors
CYPRIPEDIUM calceolus	Ladies' Slipper Orchid	Orch	Seeds	When ripe	Frame/greenhouse
			Division	March–April	Outdoors
CYSTOPTERIS	Bladder Fern	F	Spores	When ripe	Greenhouse
			Division	April	Outdoors
CYTISUS	Broom	S	Seeds	March–April	Greenhouse/frame
			Semi-ripe cuttings	Aug–Sept	Greenhouse
DABOECIA	St Dabeoc's Heath	S	Semi-ripe cuttings	June–Sept	Greenhouse/frame
			Layering (dropping)	April	Outdoors
DACTYLIS	Cocksfoot	G (P)	Seeds	May–June	Outdoors
			Division	March–April	Outdoors
DANAË	Alexandrian Laurel	S	Division	March–April	Outdoors
			Seeds	March–April	Outdoors
DAPHNE		S	Semi-ripe cuttings	Aug–Sept	Greenhouse
			Seeds	March–April	Greenhouse
			Simple layering	April–Aug	Outdoors

Plant	Type	Method	Time	Place
DARLINGTONIA California Pitcher Plant	P	Division	April	Outdoors
		Seeds	April	Greenhouse
DAVIDIA Handkerchief Tree	T	Seeds	March–April	Frame
		Semi-ripe cuttings	Aug–Sept	Greenhouse
		Air layering	April–Aug	Outdoors
DECAISNEA	S	Seeds	March–April	Frame
DELPHINIUM	P	Division	March–April	Outdoors
		Seeds	May–June	Frame/greenhouse
		Basal cuttings	April–May	Greenhouse/frame
Larkspur	An	Seeds	March–April	Outdoors
			Sept	Outdoors
DENDROMECON Tree Poppy	S	Seeds	March–April	Greenhouse
		Semi-ripe cuttings	Aug–Sept	Greenhouse
		Root cuttings	Dec–Jan	Greenhouse
DESCHAMPSIA Hair Grass	G (P)	Seeds	May–June	Outdoors
DESFONTAINEA	S	Semi-ripe cuttings	Sept–Oct	Greenhouse
		Seeds	March–April	Greenhouse
		Simple layering	April–Aug	Outdoors
DEUTZIA	S	Hardwood cuttings	Nov–Dec	Frame
		Semi-ripe cuttings	June–Aug	Greenhouse/frame
DIANTHUS Pink, Carnation	P A	Semi-ripe cuttings or pipings	July–Aug	Frame
		Layering	July–Aug	Outdoors

	Common name		Method	Time	Location
barbatus	Annual Carnation	An	Seeds	Jan–March	Greenhouse
	Sweet William	Bi		May–June	Outdoors
			Seeds	Jan–March	Greenhouse
			Seeds	May–June	Outdoors
DICENTRA		P	Division	March–April or after flowering	Outdoors
			Root cuttings	Dec–Jan	Frame
			Seeds	May–June	Frame
DICTAMNUS	Burning Bush	P	Seeds	When ripe	Outdoors
DIDISCUS		An	Seeds	March–April	Outdoors
DIERAMA	Wand Flower	Cm	Division of clumps	March–April	Outdoors
			Seeds	March–April	Frame/greenhouse
DIERVILLA		S	Hardwood cuttings	Nov–Dec	Frame
			Semi-ripe cuttings	June–Aug	Greenhouse/frame
DIGITALIS	Foxglove	Bi P	Seeds	May–June	Outdoors
DIONYSIA		A	Seeds	When ripe	Outdoors
DIPELTA		S	Semi-ripe cuttings	Aug–Sept	Greenhouse
DIPSACUS	Teasel	Bi	Seeds	May–June	Outdoors
DISANTHUS		S	Semi-ripe cuttings	Aug–Sept	Greenhouse
			Simple layering	April–Aug	Outdoors
DISPORUM		P	Division	March–April	Outdoors
			Seeds	March–April	Greenhouse

Plant		Type	Method	Time	Place
DISTYLIUM		S T	Semi-ripe cuttings	Aug–Sept	Greenhouse
DODECATHEON	Shooting Stars	P	Seeds	When ripe	Greenhouse/frame
			Division	March–April	Outdoors
DORONICUM	Leopard's Bane	P	Division	After flowering	Outdoors
DORYCNIUM		S	Seeds	March–April	Greenhouse/frame
			Softwood cuttings	June	Greenhouse
DOUGLASIA		A	Seeds	When ripe or Jan–Feb	Outdoors
DRABA	Whitlow Grass	A	Seeds	Jan–Feb	Outdoors/frame
			Division (if possible)	After flowering	Outdoors
			Cuttings (single rosettes)	Aug	Frame/greenhouse
DRACOCEPHALUM	Dragon's Head	P	Division	March–April	Outdoors
			Seeds	May–June	Outdoors
			Soft cuttings	April–June	Greenhouse
		An	Seeds	April	Outdoors
DRACUNCULUS	Dragon Plant	P	Division	March–April	Outdoors
DRIMYS		S	Semi-ripe cuttings	Aug–Sept	Frame
			Simple layering	April–Aug	Outdoors

DROSERA	Sundew	P	Division	March–April	Outdoors
			Seeds	March–April	Greenhouse
DRYAS		A	Semi-ripe cuttings	Aug	Frame
			Division	March–April	Outdoors
			Seeds	Jan–Feb	Outdoors
DRYOPTERIS	Buckler Fern	F	Division	March–April	Outdoors
			Spores	When ripe	Greenhouse
ECHINACEA		P	Division	March–April	Outdoors
			Seeds	May–June	Outdoors
			Root cuttings	Dec–Jan	Frame
ECHINOPS	Globe Thistle	P	Division (but resents disturbance)	March–April	Outdoors
			Seeds	May–June	Outdoors
			Root cuttings	Dec–Jan	Frame
ECHIUM	Viper's Bugloss	An	Seeds	March–April	Outdoors
EDRAIANTHUS		A	Seeds	Jan–Feb	Outdoors
			Softwood cuttings	April–May	Frame
ELAEAGNUS		S	Semi-ripe cuttings	Sept–Oct	Greenhouse
			Seeds	Autumn	Outdoors
ELODEA		Aq	Division	April–June	Outdoors
			Cuttings	April–June	Outdoors
ELSHOLTZIA		S	Semi-ripe cuttings	June–July	Greenhouse/frame
ELYMUS	Lyme Grass	G (P)	Division	March–April	Outdoors

Plant		Type	Method	Time	Place
EMBOTHRIUM	Chilean Fire Bush	S	Seeds	March–April	Greenhouse
			Root cuttings	Dec–Jan	Greenhouse
EMPETRUM	Crow Berry	A	Semi-ripe cuttings	June–Aug	Frame
			Layering (dropping)	April	Outdoors
ENDYMION	Bluebell	B	Seeds	Autumn or	Outdoors
				Jan–Feb	
			Bulblets	Sept–Oct	Outdoors
ENKIANTHUS		S	Seeds	March–April	Greenhouse
			Semi-ripe cuttings	Aug–Sept	Greenhouse
			Simple layering	April–Aug	Outdoors
EOMECON		P	Division	March–April	Outdoors
			Root cuttings	Dec–Jan	Frame
EPIGAEA		S	Seeds	March–April	Greenhouse
			Semi-ripe cuttings	Aug–Sept	Greenhouse
			Simple layering	April–Aug	Outdoors
EPILOBIUM	Willow Herb	P	Division	March–April	Outdoors
			Root cuttings(some)	Dec–Jan	Frame
			Seeds	May–June	Outdoors
EPIMEDIUM	Barrenwort	P	Division	After	Outdoors
				flowering	
			Seeds	When ripe	Outdoors
ERANTHIS	Winter Aconite	Tu	Division	Aug–Sept	Outdoors
			Seeds	When ripe	Outdoors

EREMURUS	Giant Asphodel	P	Division	Aug–Sept	Outdoors
			Seeds	When ripe	Outdoors/frame
ERICA	Heath, Heather	S	Semi-ripe cuttings	June–Sept	Greenhouse/frame
			Seeds	March–April	Greenhouse
			Layering (dropping)	April	Outdoors
ERIGERON	Flea Bane	P	Division	March–April	Outdoors
ERINACEA	Hedgehog Broom	S	Semi-ripe cuttings	July–Aug	Greenhouse/frame
			Seeds	March–April	Greenhouse/frame
ERINUS	Fairy Foxglove	A	Seeds	Jan–Feb	Outdoors
ERIOBOTRYA	Loquat	S T	Seeds	March–April	Greenhouse
			Semi-ripe cuttings	Aug–Sept	Greenhouse
ERITRICHIUM		A	Seeds	Jan–Feb	Greenhouse
ERODIUM	Stork's Bill	A	Seeds	Jan–Feb	Outdoors
			Semi-ripe cuttings	June–July	Greenhouse
ERYNGIUM	Sea Holly	P	Seeds	May–June	Outdoors
			Root cuttings	Dec–Jan	Frame
		Bi	Seeds	May–June	Outdoors
ERYSIMUM		P(Bi)	Seeds	May–June	Outdoors
ERYTHRONIUM	Dog's Tooth Violet	P	Seeds	Autumn or Jan–Feb	Outdoors
			Division	Aug–Sept	Outdoors

Plant	Type	Method	Time	Place
ESCALLONIA	S	Semi-ripe cuttings	July–Oct	Frame/low polythene tunnel/greenhouse
		Seeds	March–April	Greenhouse
ESCHSCHOLTZIA Californian Poppy	An	Seeds	March–April	Outdoors
			Sept–Oct	Outdoors
EUCALYPTUS Gum Tree	T	Seeds	March–April	Greenhouse
EUCOMIS Pineapple Flower	B	Bulblets	Sept–Oct	Outdoors
		Seeds	April	Greenhouse
EUCRYPHIA	T S	Simple or air layering	April–Oct	Outdoors
		Semi-ripe cuttings	June–July	Greenhouse
		Seeds	March–April	Greenhouse
EUONYMUS Spindle Tree	S T	Seeds	March–April	Outdoors
		Simple layering	April–Aug	Outdoors
		Semi-ripe cuttings	July	Frame
radicans and similar types		Semi-ripe cuttings	Sept–Oct	Frame/low polythene tunnel
	S	Division	March–April	Outdoors
EUPATORIUM Hemp Agrimony	P	Division	March–April	Outdoors
	S	Semi-ripe cuttings	Sept–Oct	Frame
EUPHORBIA Spurge	P	Division	March–April	Outdoors
		Seeds	May–June	Frame/outdoors
	S	Softwood cuttings	April–June	Greenhouse

	Snow on the Mountain	An		March—April	Outdoors
marginata	Snow on the Mountain		Seeds	March—April	Outdoors
EURYOPS		S	Soft or semi-ripe cuttings	April—July	Greenhouse/frame
EXOCHORDA		S	Softwood cuttings	May—June	Greenhouse
			Simple layering	April—Aug	Outdoors
FABIANA		S	Semi-ripe cuttings	Aug	Greenhouse
FAGUS	Beech	T	Grafting (spliced side veneer)	March	Greenhouse
			Grafting (whip-and-tongue)	March	Outdoors
species			Seeds	March—April	Outdoors
x FATSHEDERA		S	Semi-ripe cuttings	July—Sept	Greenhouse
FATSIA		S	Semi-ripe cuttings	Aug	Greenhouse
			Seeds	March—April	Greenhouse
FESTUCA	Fescue	G (P)	Division	March—April	Outdoors
			Seeds	May—June	Outdoors
FICUS carica	Fig	T	Hardwood cuttings	Dec—Jan	Greenhouse
FILIPENDULA		P	Division	March—April	Outdoors
			Seeds	When ripe	Frame/greenhouse
FITZROYA		C	Semi-ripe cuttings	Aug	Greenhouse
			Seeds	March—April	Greenhouse
FOENICULUM	Fennel	P	Division	March—April	Outdoors
			Seeds	March—April	Frame/greenhouse

Plant	Type	Method	Time	Place
FONTANESIA	S	Semi-ripe cuttings	Sept–Oct	Frame
FORSYTHIA Golden Bell	S	Semi-ripe cuttings	June–July	Frame/greenhouse
		Hardwood cuttings	Nov–Dec	Frame
		Simple layering	April–Aug	Outdoors
FOTHERGILLA	S	Simple layering	April–Aug	Outdoors
		Semi-ripe cuttings	July	Greenhouse
		Seeds	March–April	Greenhouse
FRAGARIA Strawberry	P	Layering runners	June–July	Outdoors
FRANKENIA Sea Heath	P	Division	March–April	Outdoors
		Softwood cuttings	April–June	Greenhouse
FRAXINUS Ash	T	Seeds	Autumn or March–April	Outdoors
cultivars		Grafting (whip-and-tongue)	Feb–March	Outdoors
FREMONTIA	S	Seeds	Feb	Greenhouse
		Softwood cuttings	May–June	Greenhouse
FRITILLARIA Fritillary	B	Bulblets	Sept–Oct	Outdoors
		Seeds	Autumn or Jan–Feb	Outdoors
FUCHSIA	S	Softwood or semi-ripe cuttings	April–Aug	Greenhouse/frame

Genus	Common name		Method	Time	Place
GAGEA		B	Seeds	Autumn or Jan–Feb	Outdoors
			Bulblets	Sept	Outdoors
GAILLARDIA	Blanket Flower	P	Seeds	May–June	Outdoors
			Division	March–April	Outdoors
			Root cuttings	Dec–Jan	Frame
		An	Seeds	March–April	Outdoors
GALANTHUS	Snowdrop	B	Seeds	When ripe	Outdoors
			Bulblets	After flowering	Outdoors
GALEGA	Goat's Rue	P	Division	March–April	Outdoors
			Basal cuttings	Spring	Greenhouse
			Seeds	May–June	Outdoors
GALEOBDOLON (syn. *Lamiastrum*)	Yellow Archangel	P	Division	March–April	Outdoors
GALIUM	Bedstraw	P	Division	March–April	Outdoors
GALTONIA		B	Bulblets	Sept or April	Outdoors
			Seeds	Jan–Feb	Outdoors
GARRYA		S	Semi-ripe cuttings	Aug	Greenhouse
GAULTHERIA		S	Division	March–April	Outdoors
			Semi-ripe cuttings	Aug–Sept	Greenhouse
			Seeds	March–April	Greenhouse
			Layering (dropping)	April	Outdoors

Plant		Type	Method	Time	Place
x GAULNETTYA		S	Division	March–April	Outdoors
			Semi-ripe cuttings	Aug–Sept	Greenhouse
GAURA		P (An)	Seeds	March–April	Outdoors
				Feb–March	Greenhouse
GENISTA	Broom and Gorse	S	Seeds	Feb–March	Greenhouse
			Semi-ripe cuttings	Sept–Oct	Greenhouse/frame
GENTIANA	Gentian	P A	Division	March–April	Outdoors
			Soft cuttings	April–June	Greenhouse
			Seeds	When ripe	Frame/outdoors
GERANIUM	Crane's Bill	P A	Division	March–April	Outdoors
			Root cuttings	Dec–Jan	Frame
			Seeds	May–June	Outdoors
GEUM	Avens	P	Division	March–April	Outdoors
			Seeds	May–June	Outdoors
GILIA		An	Seeds	March–April	Outdoors
GILLENIA		P	Division	March–April	Outdoors
			Seeds	May–June	Outdoors
			Basal cuttings	Spring	Greenhouse
GINKGO	Maidenhair Tree	T	Seeds	March–April	Frame/outdoors
			Simple or air layering	April–Aug	Outdoors
cultivars			Grafting (spliced side veneer)	March	Greenhouse

		An P	Seeds	March–April	Outdoors
GLAUCIUM	Horned Poppy		Seeds	March–April	Outdoors
GLECHOMA	Ground Ivy	P	Division	April	Outdoors
			Soft cuttings	April–Aug	Greenhouse/frame
			Layering (serpentine)	April–Aug	Outdoors
GLEDITSIA cultivars	Locust	T	Seeds	February	Greenhouse
			Grafting (spliced side veneer)	March	Greenhouse
			Grafting (whip-and-tongue)	March	Outdoors
GLOBULARIA	Globe Daisy	P S	Seeds	March–April	Greenhouse/frame
			Division	March–April	Outdoors
			Semi-ripe cuttings	Aug–Sept	Frame
GLYCERIA		G (P)	Division	March–April	Outdoors
GLYCYRRHIZA	Liquorice	P	Division	March–April	Outdoors
GODETIA		An	Seeds	Sept or March–April	Outdoors
GREVILLEA		S	Seeds	March–April	Greenhouse
			Semi-ripe cuttings	June–July	Greenhouse
GRISELINIA		S	Semi-ripe cuttings	Sept–Oct	Frame/low polythene tunnel
			Seeds	March–April	Greenhouse
GUNNERA		P	Division	April	Outdoors
			Seeds	March–April	Greenhouse
			'Eyes' of trunks	Spring	Greenhouse

Plant		Type	Method	Time	Place
GYMNOCLADUS	Kentucky Coffee Tree	T	Seeds	Feb–March	Greenhouse
			Root cuttings	Dec–Jan	Greenhouse
GYPSOPHILA		P	Seeds	May–June	Outdoors
			Soft basal cuttings	April–May	Greenhouse
		A	Seeds	Jan–Feb	Outdoors
		An	Seeds	March–April	Outdoors
HABERLEA		A	Seeds	When ripe	Frame
			Leaf cuttings	June–Aug	Frame/greenhouse
HACQUETIA		P	Division	March–April	Outdoors
			Seeds	When ripe	Outdoors
HALESIA	Snowdrop Tree	T	Seeds	March–April	Frame/greenhouse
			Simple layering	April–Aug	Outdoors
			Softwood cuttings	May–June	Greenhouse
x HALIMIOCISTUS		S	Semi-ripe cuttings	June–July	Greenhouse
				July–Aug	Frame
HALIMIUM		S	Semi-ripe cuttings	June–July	Greenhouse
				July–Aug	Frame
HAMAMELIS	Witch Hazel	S T	Simple or air layering	April–Aug	Outdoors
			Grafting (spliced side veneer)	March	Greenhouse
			Seeds	March–April	Greenhouse

Genus	Common name	Type	Method	Time	Location
HAPLOPAPPUS		S	Soft or semi-ripe cuttings	May–July	Greenhouse
		P	Division	March–April	Outdoors
HEBE	Shrubby Veronica	S	Softwood cuttings	April–June	Greenhouse
			Semi-ripe cuttings	July–Oct	Frame/low polythene tunnel
HEDERA	Ivy	Cl	Leaf-bud cuttings	June–July	Greenhouse/frame
			Layering (serpentine)	April–Aug	Outdoors
HELENIUM	Sneezeweed	P	Division	March–April	Outdoors
HELIANTHEMUM	Rock Rose	S	Semi-ripe cuttings	June–July	Greenhouse
				July	Frame
			Seeds	March–April	Greenhouse
HELIANTHUS	Sunflower	P	Division	March–April	Outdoors
		An	Seeds	March–April	Outdoors
HELICHRYSUM	Straw Flower	An	Seeds	March–April	Outdoors
		P A	Semi-ripe cuttings	June–Aug	Greenhouse
			Division	March–April	Outdoors
		S	Semi-ripe cuttings	June–Aug	Greenhouse
HELICTOTRICHON		G (P)	Division	March–April	Outdoors
			Seeds	May–June	Outdoors
HELIOPSIS		P	Division	March–April	Outdoors
			Basal cuttings	Spring	Greenhouse
			Seeds	May–June	Outdoors

217

Plant	Type	Method	Time	Place
HELIPTERUM	An	Seeds	March–April	Outdoors
HELLEBORUS Hellebore, Lenten Rose, Christmas Rose	P	Division (resents disturbance)	After flowering	Outdoors
		Seeds	When ripe	Outdoors
HEMEROCALLIS Day Lily	P	Division	March–April	Outdoors
		Seeds	May–June	Outdoors/frame
HEPATICA	P	Division	After flowering	Outdoors
		Seeds	April	Outdoors
HESPERIS Sweet Rocket	P	Division	March–April	Outdoors
		Seeds	May–June	Outdoors
HEUCHERA Alum Root	P	Division	March–April	Outdoors
		Seeds	May–June	Outdoors
x HEUCHERELLA	P	Division	March–April	Outdoors
HIBISCUS Shrubby Mallow	S	Grafting (spliced side)	March	Greenhouse
		Semi-ripe cuttings	Aug–Sept	Greenhouse
		Seeds	March–April	Outdoors
		Simple layering	April–Aug	Outdoors
	An	Seeds	March–April	Outdoors
HIERACEUM Hawkweed	P	Division	March–April	Outdoors
		Seeds	May–June	Outdoors

HIPPOPHAE	Sea Buckthorn	S	Seeds	March–April	Outdoors
			Suckers	March	Outdoors
HOHERIA	Ribbon Wood	S T	Simple or air layering	April–Aug	Outdoors
			Semi-ripe cuttings	Aug–Sept	Greenhouse
HOLCUS		G (P)	Division	March–April	Outdoors
HOLODISCUS		S	Semi-ripe cuttings	Aug–Sept	Greenhouse/frame
			Hardwood cuttings	Nov	Frame
HORDEUM	Foxtail Barley or Squirrel Tail Grass	G (An)	Seeds	March–April	Outdoors
HORMINUM		P	Division	March–April	Outdoors
			Seeds	May–June	Outdoors
HOSTA	Plantain Lily	P	Division	March–April	Outdoors
			Seeds	April–May	Frame
HOTTONIA	Water Violet	Aq	Division	April–June	Outdoors
HOUSTONIA	Bluets	A	Division	March–April	Outdoors
HOUTTUYNIA		P	Division	March–April	Outdoors
			Root cuttings	Dec–Jan	Frame
HUMULUS	Hop	Cl	Seeds	April	Frame
			Root cuttings	Dec–Jan	Frame
			Division	March–April	Outdoors
HUTCHINSIA		A	Division	March–April	Outdoors
			Seeds	Jan–Feb	Outdoors

Plant		Type	Method	Time	Place
HYACINTHUS	Hyacinth	B	Bulblets	Sept–Oct	Outdoors
			Seeds	When ripe or	Outdoors
				Jan–Feb	
HYDRANGEA		S	Seeds	March–April	Greenhouse
			Semi-ripe cuttings	July–Aug	Greenhouse/frame
				Aug–Sept	Outdoors
petiolaris	Climbing Hydrangea	Cl	Seeds	March–April	Greenhouse
			Layering (serpentine)	April–Aug	Outdoors
HYDROCHARIS	Frogbit	Aq	Rooted runners	April–June	Outdoors
HYMENANTHERA		S	Semi-ripe cuttings	July–Oct	Frame
			Seeds	March–April	Greenhouse
HYPERICUM	St John's Wort	S	Semi-ripe cuttings	Sept–Oct	Frame/low polythene tunnel
			Division (where possible)	March–April	Outdoors
			Seeds	March–April	Greenhouse
		P A	Division	March–April	Outdoors
			Soft or semi-ripe cuttings	June–July	Greenhouse/frame
			Seeds	March–April	Greenhouse/frame
HYPOXIS		P (Cm)	Division of clumps	When dormant	Outdoors/ greenhouse

			Method	Time	Location
HYSSOPUS	Hyssop	P	Seeds	April	Outdoors
			Division	March–April	Outdoors
			Soft cuttings	April–May	Frame/greenhouse
			Layering (dropping)	April–May	Outdoors
IBERIS	Perennial Candytuft	P A	Soft or semi-ripe cuttings	June–July	Greenhouse/frame
	Candytuft	An	Seeds	March–April	Outdoors
IDESIA		T	Seeds	March–April	Greenhouse
			Air or simple layering	April–Aug	Outdoors
ILEX	Holly	T S	Semi-ripe cuttings	Oct–Dec	Greenhouse/frame
			Seeds	March–April	Outdoors
			Simple layering	April–Aug	Outdoors
INCARVILLEA		P	Seeds	March–April	Greenhouse
			Division	March–April	Outdoors
INDIGOFERA		S	Seeds	March–April	Greenhouse
			Semi-ripe cuttings	July–Aug	Greenhouse
INULA		P	Division	March–April	Outdoors
			Seeds	May–June	Outdoors
IONOPSIDIUM	Violet Cress	An	Seeds	March–April	Outdoors
IPHEION		B	Bulblets	Sept	Outdoors
			Seeds	Jan–Feb	Outdoors

Plant		Type	Method	Time	Place
IRIS					
rhizomatous species	e.g. Bearded or German Iris, Water Iris	P	Division	After flowering	Outdoors
			Seeds	When ripe	Outdoors
bulbous species	e.g. Dwarf alpine, Dutch, Spanish, English	B	Bulblets	Sept—Oct	Outdoors
			Seeds	When ripe or Jan—Feb	Outdoors
ISATIS	Woad	Bi	Seeds	May—June	Outdoors
ITEA		S	Semi-ripe cuttings	Aug—Sept	Frame/greenhouse
			Simple layering	April—Aug	Outdoors
IXIA	Corn Lily	Cm	Cormlets	Oct—Nov or spring	Outdoors
			Seeds	Autumn	Outdoors
IXIOLIRION		B	Bulblets	Oct or April	Outdoors
JASMINUM	Jasmine	Cl	Semi-ripe cuttings	July—Aug	Frame/greenhouse
			Hardwood cuttings	Nov	Frame
			Serpentine layering	April—Aug	Outdoors
JEFFERSONIA		A	Division	March—April	Outdoors
			Seeds	When ripe	Outdoors
JUGLANS	Walnut	T	Seeds	March—April	Outdoors
			Grafting (whip-and-tongue)	March	Outdoors

			Method	Time	Location
JUNCUS	Rush	P	Division	April–June	Outdoors
JUNIPERUS	Juniper	C	Semi-ripe cuttings	Sept–Jan	Greenhouse/frame
			Seeds	March–April	Outdoors
			Simple layering	April–Aug	Outdoors
			Grafting (veneer)	March	Greenhouse
KALMIA		S	Simple layering (difficult)	April–Aug	Outdoors
			Semi-ripe cuttings (difficult)	Aug–Sept	Greenhouse
			Seeds	March–April	Greenhouse
KERRIA	Jew's Mallow	S	Hardwood cuttings	Oct–Nov	Frame
			Semi-ripe cuttings	June–July	Frame
			Simple layering	April–Aug	Outdoors
			Division	March	Outdoors
KIRENGESHOMA		P	Division	March–April	Outdoors
			Seeds	March–April	Greenhouse
KNIPHOFIA	Red-Hot Poker	P	Division	March–April	Outdoors
			Seeds	May–June	Outdoors
KOELREUTERIA	Golden Rain Tree	T	Seeds	March–April	Greenhouse
			Root cuttings	Dec–Jan	Greenhouse
KOLKWITZIA	Beauty Bush	S	Semi-ripe cuttings	July–Aug	Greenhouse/frame
+ LABURNOCYTISUS		T	Grafting (whip-and-tongue) (Laburnum stocks)	Feb–March	Outdoors

Plant		Type	Method	Time	Place
LABURNUM		T	Seeds	March–April	Outdoors
			Grafting (whip-and-tongue)	March	Outdoors
			Budding	July	Outdoors
			Hardwood cuttings	Nov–Dec	Greenhouse/outdoors
LACTUCA		P	Seeds	May–June	Outdoors
			Division	March–April	Outdoors
LAGURUS	Hare's Tail Grass	G (An)	Seeds	March–April	Outdoors
				Aug–Sept	Frame
LAMARCKIA		G (An)	Seeds	April–May	Outdoors
				Aug–Sept	Frame
LAMIUM		P	Division	March–April	Outdoors
			Seeds	May–June	Outdoors
LARIX	Larch	C	Seeds	March–April	Outdoors
			Grafting (veneer)	March	Greenhouse
LATHYRUS	Sweet Pea	Cl An	Seeds	Sept–Oct	Frame
				March–April	Outdoors
		P	Division	March–April	Outdoors
			Seeds	March–April	Frame/greenhouse
LAURUS	Bay	T	Semi-ripe cuttings	Sept–Oct	Frame/low polythene tunnel
			Seeds	March–April	Outdoors

					Frame/low polythene tunnel
LAVANDULA	Lavender	S	Semi-ripe cuttings	Sept–Oct	Outdoors
			Layering (dropping)	April	Outdoors
LAVATERA	Mallow	P	Division	March–April	Outdoors
			Semi-ripe cuttings	July–Aug	Greenhouse
		An	Seeds	March–April	Outdoors
LAYIA	Tidy Tips	An	Seeds	March–April	Outdoors
LEDUM		S	Seeds	March–April	Greenhouse
			Simple layering	April–Aug	Outdoors
			Division	March–April	Outdoors
			Semi-ripe cuttings	July–Aug	Greenhouse
LEIOPHYLLUM	Sand Myrtle	S	Semi-ripe cuttings	July–Aug	Greenhouse
			Simple layering	April–Aug	Outdoors
LEMNA	Duckweed	Aq	Division	April–June or any time	Outdoors
LEONTOPODIUM	Edelweiss	A	Seeds	Jan–Feb	Outdoors
			Division	March–April	Outdoors
LEPTOSPERMUM		S T	Semi-ripe cuttings	Aug	Greenhouse
			Seeds	March–April	Greenhouse
LESPEDESA	Bush Clover	S	Seeds	March–April	Greenhouse
			Basal cuttings	Spring	Greenhouse
			Division	March–April	Outdoors

225

Plant		Type	Method	Time	Place
LEUCOJUM	Snowflake	B	Bulblets	Sept	Outdoors
			Seeds	When ripe or Jan–Feb	Outdoors
autumnale			Bulblets	Summer	Outdoors
LEUCOTHOË		S	Seeds	March–April	Greenhouse
			Semi-ripe cuttings	Aug	Greenhouse/frame
			Simple layering	April–Aug	Outdoors
LEVISTICUM	Lovage	P	Division	March–April	Outdoors
			Seeds	May–June	Outdoors
LEWISIA		A	Division	March–April	Outdoors
			Leaf cuttings	June–Aug	Frame/greenhouse
			Seeds	When ripe	Outdoors
LEYCESTERIA	Himalaya Honeysuckle	S	Semi-ripe cuttings	Aug–Sept	Frame
			Hardwood cuttings	Nov–Dec	Frame
			Seeds	March–April	Greenhouse
LIATRIS		P	Division	March–April	Outdoors
			Seeds	May–June	Outdoors
LIBERTIA		P	Division	March–April	Outdoors
			Seeds	March–April	Greenhouse
LIGULARIA		P	Division	March–April	Outdoors
			Seeds	May–June	Outdoors

LIGUSTRUM	Privet	**S**			
ovalifolium and vulgare other species			Hardwood cuttings	Nov–Dec	Outdoors
			Semi-ripe cuttings	July–Aug	Greenhouse/frame
			Simple layering	April–Aug	Outdoors
all species			Seeds	March–April	Outdoors
LILIUM	Lily	**B**	Bulblets	Autumn	Outdoors
			Bulbils	Spring	Outdoors
			Scales	Summer or spring	Greenhouse/frame
			Seeds	When ripe	Frame
LIMNANTHES	Poached Egg Flower	**An**	Seeds	March–April	Outdoors
				Sept	Outdoors
LIMONIUM	Sea Lavender	**An**	Seeds	Jan–March	Greenhouse
				April–May	Outdoors
		P	Division	March–April	Outdoors
			Seeds	May–June	Outdoors/frame
latifolium		**P**	Root cuttings	Dec–Jan	Frame
LINANTHUS		**An**	Seeds	March–April	Outdoors
				Sept	Outdoors
LINARIA	Toadflax	**An**	Seeds	March–April	Outdoors
				Sept	Outdoors
		P	Soft cuttings	April–June	Greenhouse
			Seeds	March–April	Greenhouse
			Division	March–April	Outdoors

Plant		Type	Method	Time	Place
LINUM	Flax	An	Seeds	March–April	Outdoors
		P A	Seeds	March–April	Greenhouse/frame
			Soft cuttings	May–June	Greenhouse
			Semi-ripe cuttings	July–Aug	Frame/greenhouse
		S	Seeds	March–April	Greenhouse
			Softwood cuttings	May–June	Greenhouse
LIPPIA	Lemon Verbena	S	Softwood cuttings	April–June	Greenhouse
LIQUIDAMBAR	Sweet Gum	T	Seeds	March–April	Outdoors
			Air or simple layering	April–Aug	Outdoors
LIRIODENDRON cultivars	Tulip Tree	T	Seeds	March–April	Outdoors
			Grafting (spliced side veneer)	March	Greenhouse
LIRIOPE		P	Division	March–April	Outdoors
			Seeds	May–June	Outdoors
LITHOSPERMUM	Gromwell	S	Semi-ripe cuttings	July	Greenhouse
			Seeds	March–April	Greenhouse
LOBELIA		P	Division	March–April	Outdoors
		S	Soft or semi-ripe cuttings	April–Aug	Greenhouse
		P S	Seeds	March–April	Greenhouse
LOBULARIA	Sweet Alyssum	An	Seeds	Feb–March	Greenhouse
				April–May	Outdoors

LOMATIA		T S	Seeds	March–April	Greenhouse
			Semi-ripe cuttings	Aug	Greenhouse
LONAS	African Daisy	An	Seeds	March–April	Outdoors
LONICERA	Honeysuckle	Cl	Semi-ripe cuttings	June–July	Greenhouse
				Oct–Nov	Frame
			Serpentine layering	April–Aug	Outdoors
		S	Semi-ripe cuttings	July–Aug	Greenhouse/frame
			Hardwood cuttings	Nov–Dec	Frame/outdoors
		Cl S	Seeds	March–April	Greenhouse
LOROPETALUM		S	Seeds	March–April	Greenhouse
			Simple layering	April–Aug	Outdoors
			Semi-ripe cuttings	July–Aug	Greenhouse
LOTUS		P	Seeds	March–April	Greenhouse
			Division	March–April	Outdoors
			Semi-ripe cuttings	June–Aug	Greenhouse
LUNARIA	Honesty				
annua		An Bi	Seeds	May–June	Outdoors
rediviva		P	Seeds	May–June	Outdoors
			Division	March–April	Outdoors
LUPINUS	Lupin	P	Soft basal cuttings	April–May	Frame
			Seeds	May–June	Frame/greenhouse
		An	Seeds	March–April	Outdoors
arboreus	Tree Lupin	S	Seeds	Jan–Feb	Greenhouse/frame
			Semi-ripe cuttings	July–Aug	Greenhouse

Plant		Type	Method	Time	Place
LUZULA	Woodrush	P	Division	March–April	Outdoors
LYCHNIS	Campion	P	Seeds	May–June	Outdoors
			Division	March–April	Outdoors
LYCIUM	Box Thorn	S	Seeds	March–April	Greenhouse
			Semi-ripe cuttings	July–Aug	Greenhouse
			Suckers	March–April	Outdoors
LYONIA		S	Seeds	March–April	Greenhouse
			Simple layering	April–Aug	Outdoors
			Semi-ripe cuttings	Aug	Greenhouse
LYSICHITUM	Skunk Cabbage	P	Division	After flowering or autumn	Outdoors
			Seeds	When ripe	Frame/greenhouse
LYSIMACHIA	Loosestrife	P	Division	March–April	Outdoors
			Seeds	May–June	Outdoors
			Basal cuttings	Spring	Greenhouse
LYTHRUM	Purple Loosestrife	P	Division	March–April	Outdoors
			Seeds	May–June	Outdoors
			Basal cuttings	Spring	Greenhouse
MAACKIA		S T	Seeds	March–April	Greenhouse
			Root cuttings	Dec–Jan	Greenhouse

			Method	Time	Location
MACLEAYA	Plume Poppy	P	Root cuttings	Dec–Jan	Greenhouse
			Division	March–April	Outdoors
			Basal cuttings	June	Greenhouse
			Suckers	Summer	Outdoors
MAGNOLIA		T S	Simple layering or air layering	April–Aug	Outdoors
			Seeds	Autumn or spring	Frame/greenhouse
grandiflora			Softwood cuttings	April–June	Greenhouse
			Semi-ripe cuttings	Aug	Greenhouse
MAHONIA		S	Seeds	When ripe or March–April	Greenhouse/frame
aquifolium	Oregon Grape		Leaf-bud cuttings	Oct or April	Greenhouse
			Division	March–April	Outdoors
			Seeds	When ripe or March–April	Frame/greenhouse
			Leaf-bud cuttings	Oct or April	Greenhouse
MALCOMIA	Virginian Stock	An	Seeds	March–April	Outdoors
				Sept	Outdoors
MALOPE		An	Seeds	March–April	Outdoors
MALUS	Apples, Crab Apples	T	Grafting (whip-and-tongue)	March	Outdoors
species			Budding	June–July	Outdoors
			Seeds	March–April	Outdoors
MALVA	Mallow	P	Soft cuttings	April–June	Greenhouse
			Seeds	March–April	Greenhouse

Plant		Type	Method	Time	Place
MARGYRICARPUS	Pearl Fruit	S	Semi-ripe cuttings	July–Aug	Greenhouse
			Simple layering	April–Aug	Outdoors
			Seeds	March–April	Frame
MATTEUCCIA	Ostrich Fern	F	Division	March–April	Outdoors
			Spores	When ripe	Greenhouse
MATTHIOLA bicornis	Stock				
	Night-scented Stock	An	Seeds	March–April	Outdoors
	Brompton Stock	Bi	Seeds	June–July	Frame
	East Lothian Stock	Bi	Seeds	Feb–March	Greenhouse
				July–Aug	Frame
	Beauty of Nice Stock	An	Seeds	Feb–March	Greenhouse
				July–Aug	Frame
	Ten-week Stock	An	Seeds	Feb–March	Greenhouse
MAZUS		A	Division	March–April	Outdoors
MECONOPSIS	including Blue and Welsh Poppies	P An Bi	Seeds	When ripe	Frame
		P	Division	March–April	Outdoors
MELANDRIUM	Campion	P	Seeds	May–June	Outdoors
			Division	March–April	Outdoors
MELISSA	Lemon Balm	P	Division	March–April	Outdoors
			Seeds	May–June	Outdoors
MELITTIS	Bastard Balm	P	Division	March–April	Outdoors
			Seeds	May–June	Outdoors

MENTHA	Mint	P A	Division	March–April	Outdoors
			Soft cuttings	April–June	Frame/greenhouse
MENTZELIA (Bartonia)		An	Seeds	March–April	Outdoors
MENYANTHES	Bog Bean	Aq	Division	April–June	Outdoors
MENZIESIA		S	Simple layering	April–Aug	Outdoors
			Semi-ripe cuttings	July–Aug	Greenhouse
			Seeds	March–April	Greenhouse
MERTENSIA	Smooth Lungwort	P	Division	March–April	Outdoors
			Seeds	When ripe	Frame
MESPILUS	Medlar	T	Seeds	March–April	Outdoors
			Simple or air layering	April–Aug	Outdoors
METASEQUOIA	Dawn Redwood	C	Hardwood cuttings	Nov–Dec	Greenhouse
			Seeds	March–April	Greenhouse/frame
MICHELIA		T	Simple or air layering	April–Aug	Outdoors
			Semi-ripe cuttings	Aug–Sept	Greenhouse
MICROMERIA		S P	Semi-ripe cuttings	Aug–Sept	Frame/greenhouse
			Seeds	March–April	Frame/greenhouse
			Division	March–April	Outdoors
MILIUM	Millet	G (P)	Division	March–April	Outdoors
			Seeds	May–June	Outdoors

Plant		Type	Method	Time	Place
MIMULUS	Monkey Flower	P S	Seeds	Feb–March	Greenhouse
			Soft cuttings	April–June	Greenhouse
			Division	March–April	Outdoors
MINUARTIA		A	Division	March–April	Outdoors
MISCANTHUS		G (P)	Division	March–April	Outdoors
MITCHELLA	Partridge Berry	S	Division	March–April	Outdoors
			Simple layering	April–Aug	Outdoors
			Semi-ripe cuttings	July–Aug	Greenhouse
MITELLA	Mitrewort	P	Division	March–April	Outdoors
			Seeds	March–April	Greenhouse/frame
MOLINIA		G (P)	Division	March–April	Outdoors
			Seeds	May–June	Outdoors
MOLTKIA		P	Semi-ripe cuttings	July	Greenhouse
MOLUCCELLA	Bells of Ireland	An	Seeds	March–April	Outdoors
				Feb–March	Greenhouse
MONARDA	Bee Balm, Bergamot or Oswego Tea	P	Division	March–April	Outdoors
MONTIA		P An	Seeds	When ripe	Frame
			Division	March–April	Outdoors
MORINA		P	Seeds	May–June	Outdoors
			Division	March–April	Outdoors
MORISIA		A	Root cuttings	Dec–Jan	Frame/greenhouse
			Seeds	Jan–Feb	Outdoors

MORUS	Mulberry	T	Hardwood cuttings	Nov–Dec	Outdoors
			Simple layering	April–Aug	Outdoors
MUEHLENBECKIA	Wire Plant	S	Semi-ripe cuttings	Oct–Nov	Frame/low polythene tunnel
			Semi-ripe cuttings	July–Aug	Frame
			Division	March	Outdoors
MUSCARI	Grape Hyacinth	B	Bulblets	Sept	Outdoors
			Seeds	When ripe or Jan–Feb	Outdoors
MUTISIA		S Cl	Seeds	March–April	Greenhouse
			Semi-ripe cuttings	July–Aug	Greenhouse
MYOSOTIDIUM		P	Seeds	March–April	Greenhouse
			Division	After flowering	Outdoors
MYOSOTIS	Forget-me-not including Water Forget-me-nots	Bi	Seeds	May–June	Outdoors
		P	Seeds	May–June	Outdoors
			Division	March–April or after flowering	Outdoors
			Soft cuttings	June–Aug	Frame/greenhouse
MYRICA	Wax and Bog Myrtles	S	Seeds	When ripe	Greenhouse
			Semi-ripe cuttings	July–Aug	Greenhouse
			Division	March–April	Outdoors
MYRIOPHYLLUM	Water Milfoil	Aq	Soft cuttings	April–June	Outdoors
			Division	April–June	Outdoors
MYRRHIS	Myrrh	P	Seeds	When ripe	Outdoors
			Division	March–April	Outdoors

Plant		Type	Method	Time	Place
MYRTUS	Myrtle	S	Seeds	March–April	Greenhouse
			Semi-ripe cuttings	July–Aug	Greenhouse
NANDINA	Heavenly Bamboo	S	Semi-ripe cuttings	July–Aug	Frame/greenhouse
			Seeds	When ripe	Frame
NARCISSUS	Daffodil	B	Bulblets	Sept	Outdoors
			Seeds	When ripe or Jan–Feb	Outdoors
NEILLIA		S	Semi-ripe cuttings	July–Aug	Greenhouse
			Hardwood cuttings	Nov	Frame
NEMOPHILA	Baby Blue Eyes	An	Seeds	March–April	Outdoors
NEPETA	Catmint	P	Division	March–April	Outdoors
			Soft cuttings	April–June	Greenhouse
			Seeds	May–June	Outdoors
NERINE	including Guernsey Lily	B	Bulblets	Aug	Outdoors
			Seeds	When ripe	Greenhouse/frame
NERTERA		P	Division	March–April	Outdoors
			Seeds	March–April	Greenhouse/frame
NICANDRA	Shoo-fly, Apple of Peru	An	Seeds	April–May	Outdoors
				Feb–March	Greenhouse
NIEREMBERGIA repens	Cup Flower	S	Semi-ripe cuttings	July–Aug	Greenhouse
		P	Division	March–April	Outdoors
NIGELLA	Love-in-a-Mist	An	Seeds	March–April	Outdoors
				Sept–Oct	Outdoors

NOMOCHARIS		B	Seeds	Feb	Greenhouse/frame
			Scales	Spring or summer	Greenhouse
NOTHOFAGUS	Southern Beech	T	Seeds	March–April	Outdoors
			Simple or air layering	April–Aug	Outdoors
			Semi-ripe cuttings	July–Aug	Greenhouse
NOTHOLIRION		B	Bulblets	After flowering	Outdoors
NOTOSPARTIUM	Southern Broom	S	Seeds	March–April	Greenhouse/frame
NUPHAR	Yellow Water Lily	Aq	Division of tubers	April–June	Outdoors
			Seeds	When ripe	Frame
NYMPHAEA	Water Lily	Aq	Division of tubers	April–June	Outdoors
			Seeds	When ripe	Frame
NYMPHOIDES	Fringed Water Lily	Aq	Cuttings of runners	April–June	Outdoors
NYSSA	Tupelo Tree	T	Seeds (pots)	Feb–March	Greenhouse
			Simple layering into pots	April–Aug	Outdoors
			Air layering	April–Aug	Outdoors
OENOTHERA	Evening Primrose	An	Seeds	Feb–March	Greenhouse
		P	Seeds	Feb–March	Greenhouse
			Division	March–April	Outdoors
			Soft cuttings	April–June	Frame
		Bi	Seeds	July	Frame

237

Plant		Type	Method	Time	Place
OLEARIA	Daisy Bush	S	Semi-ripe cuttings	Aug–Sept	Frame/low polythene tunnel
OMPHALODES	Navelwort	P A	Division Seeds	March–April March–April	Outdoors Frame
ONOCLEA	Sensitive Fern	F	Division Spores	March–April When ripe	Outdoors Greenhouse
ONONIS		P S An	Soft cuttings Semi-ripe cuttings Seeds Seeds	May–June July–Aug March–April March–April	Greenhouse Greenhouse Greenhouse Outdoors
ONOPORDON	Scotch Thistle	Bi	Seeds	May–June	Outdoors
ONOSMA		A	Seeds Semi-ripe cuttings	March–April June–Aug	Greenhouse Greenhouse
OPHIOPOGON		P	Division	March–April	Outdoors
OPHRYS	including Bee Orchid	Orch	Division Seeds	Autumn When ripe	Outdoors Greenhouse
ORCHIS	including Spotted, Pyramidal and Marsh Orchid	Orch	Division Seeds	Autmn When ripe	Outdoors Greenhouse
ORIGANUM	Marjoram	P	Division Seeds Soft basal cuttings	March–April Feb–March May–June	Outdoors Greenhouse Greenhouse

Genus	Common name		Method	Time	Location
ORNITHOGALUM		B	Bulblets Seeds	Sept When ripe or Jan–Feb	Outdoors Outdoors/frame
ORONTIUM	Golden Club	Aq	Division Seeds	April–June When ripe	Outdoors Frame
OSMANTHUS		S	Semi-ripe cuttings Simple layering	Sept–Oct April–Aug	Greenhouse/frame Outdoors
× OSMAREA		S	Semi-ripe cuttings	Sept–Oct	Greenhouse/frame
OSMARONIA	Indian Plum, Oso Berry	S	Suckers	Nov–March	Outdoors
OSMUNDA	including Royal Fern	F	Division Spores	March–April When ripe	Outdoors Greenhouse
OSTEOMELES		S	Seeds Semi-ripe cuttings	March–April July–Aug	Greenhouse Greenhouse
OSTROWSKIA		P	Seeds Root cuttings	When ripe Dec–Jan	Frame/outdoors Frame
OSTRYA	Hop Hornbeam	T	Seeds	March–April	Outdoors
OTHONNOPSIS		S	Semi-ripe cuttings Division	June–July March–April	Greenhouse/frame Outdoors
OURISIA	Mountain Foxglove	P	Seeds Division	March–April March–April	Greenhouse/frame Outdoors
OXALIS	Wood Sorrel	P A	Division Seeds	March–April March–April	Outdoors Greenhouse/frame

239

Plant		Type	Method	Time	Place
OXYDENDRUM	Sorrel Tree	T	Seeds	March–April	Greenhouse
			Semi-ripe cuttings	Aug–Sept	Greenhouse
			Simple or air layering	April–Aug	Outdoors
OZOTHAMNUS		S	Semi-ripe cuttings	July–Aug	Frame/greenhouse
PACHYSANDRA	Mountain Spurge	P	Division	March–April	Outdoors
			Semi-ripe cuttings	Aug–Sept	Frame
PAEONIA	Peony	P	Division (resents root disturbance)	March–April	Outdoors
	Tree Peony	S	Seeds	When ripe	Greenhouse
			Simple layering	March–April	Outdoors
PALIURUS		S	Seeds	When ripe	Greenhouse/frame
			Root cuttings	Dec–Jan	Greenhouse/frame
PANCRATIUM		B	Bulblets	Aug–Sept	Outdoors
			Seeds	When ripe or Jan–Feb	Outdoors/frame
PANICUM		G (P)	Division	March–April	Outdoors
		An	Seeds	March–June	Outdoors
PAPAVER	Poppy				
	including Shirley Poppy	An	Seeds	March–April	Outdoors
	including Iceland Poppy	P	Seeds	May–June	Outdoors
			Division	March–April	Outdoors
orientale	Oriental Poppy	P	Root cuttings	Dec–Jan	Frame

Genus	Common name	Type	Method	Timing	Location
PARAQUILEGIA		A	Seeds	When ripe or Jan–Feb	Outdoors
			Division	March–April	Outdoors
PAROCHETUS	Shamrock Pea, Blue Oxalis	P	Division	March–April	Outdoors
			Seeds	March–April	Greenhouse
PARROTIA	Iron Tree	T	Simple layering	April–Aug	Outdoors
			Seeds	When ripe	Frame
PARTHENOCISSUS	including Virginia Creeper	Cl	Softwood cuttings	April–June	Greenhouse
			Hardwood cuttings	Nov–Dec	Greenhouse
			Serpentine layering	April–Aug	Outdoors
			Seeds	March–April	Outdoors
PASSIFLORA	Passion Flower	Cl	Leaf-bud cuttings	May–June	Greenhouse
			Serpentine layering	April–Aug	Outdoors
			Seeds	March–April	Greenhouse
PAULOWNIA		T	Root cuttings	Dec–Jan	Greenhouse
			Seeds	March–April	Greenhouse
PELTIPHYLLUM	Umbrella Plant	P	Division	March–April	Outdoors
			Seeds	March–April	Frame
PENNISETUM		G (An P)	Seeds	Feb–March	Greenhouse
			Division	March–April	Outdoors
PENSTEMON	Beard Tongue	P A S	Softwood cuttings	May–June	Frame
			Semi-ripe cuttings	Aug–Sept	Frame
			Division	March–April	Outdoors
			Seeds	Feb–March	Greenhouse

Plant	Type	Method	Time	Place
PERIPLOCA Silk Vine	Cl	Semi-ripe cuttings	July–Aug	Greenhouse
		Simple or serpentine layering	Aug–Sept	Outdoors
		Division	March–April	Outdoors
PERNETTYA Prickly Heath	S	Division	March–April	Outdoors
		Layering (dropping)	April	Outdoors
		Semi-ripe cuttings	July–Aug	Greenhouse/frame
		Seeds	March–April	Greenhouse
PEROVSKIA	P	Semi-ripe cuttings	July–Aug	Greenhouse/frame
		Hardwood cuttings	Nov	Frame/outdoors
PETASITES Butter Bur, Winter Heliotrope	P	Division	March–April or after flowering	Outdoors
PETROPHYTUM	A	Soft cuttings	After flowering	Greenhouse
		Seeds	When ripe or Jan–Feb	Greenhouse
PHACELIA	An	Seeds	March–April	Outdoors
PHALARIS Ribbon Grass	G (P)	Division	March–April	Outdoors
	G (An)	Seeds	March–April	Outdoors
PHELLODENDRON Cork Tree	T	Seeds	March–April	Greenhouse
		Semi-ripe cuttings	July	Greenhouse

			Method	Time	Location
PHILADELPHUS	Mock Orange	S	Softwood cuttings	May–June	Greenhouse
			Semi-ripe cuttings	July–Aug	Frame
			Hardwood cuttings	Dec–Jan	Frame
			Seeds	March–April	Greenhouse
PHILLYREA	Jasmine Box	S T	Semi-ripe cuttings	Aug	Frame
			Simple layering	April–Aug	Outdoors
PHLOMIS		P	Seeds	March–April	Greenhouse
			Division	March–April	Outdoors
		S	Seeds	March–April	Greenhouse
			Semi-ripe cuttings	July–Aug	Greenhouse/frame
PHLOX paniculata	Border Phlox	P	Root cuttings	Dec–Jan	Greenhouse/frame
			Division	March–April	Outdoors
		A	Semi-ripe cuttings	June–July	Greenhouse/frame
			Seeds	Jan–Feb	Outdoors
PHORMIUM	New Zealand Flax	P	Division	March–April	Outdoors
			Seeds	March–April	Greenhouse
PHOTINIA		T S	Seeds	March–April	Outdoors
			Semi-ripe cuttings	July–Aug	Greenhouse
			Semi-ripe cuttings	Sept–Oct	Frame
			Simple layering	April–Aug	Outdoors
PHUOPSIS		P	Division	March–April	Outdoors
			Basal cuttings	July–Aug	Frame/greenhouse
PHYGELIUS	Cape Figwort	S	Seeds	March–April	Greenhouse
			Semi-ripe cuttings	June–Aug	Greenhouse
			Division	March–April	Outdoors

Plant		Type	Method	Time	Place
PHYLLITIS	Hart's Tongue Fern	F	Division	March–April	Outdoors
			Spores	When ripe	Greenhouse
PHYLLODOCE	Mountain Heath	S	Seeds	March–April	Greenhouse/frame
			Semi-ripe cuttings	July–Aug	Greenhouse/frame
			Simple layering	April–Aug	Outdoors
PHYLLOSTACHYS	Bamboo	G (P)	Division	March–April	Outdoors
PHYSALIS	Chinese Lantern	P	Division	March–April	Outdoors
			Root cuttings	Dec–Jan	Frame
PHYSOCARPUS		S	Seeds	March–April	Outdoors
			Semi-ripe cuttings	June–July	Frame
			Hardwood cuttings	Nov–Dec	Frame
PHYSOSTEGIA	False Dragonhead	P	Division	March–April	Outdoors
PHYTEUMA	Horned Rampion	A	Seeds	Jan–Feb	Frame
			Division	March–April	Outdoors
PHYTOLACCA	Pokeweed	P	Division	March–April	Outdoors
			Seeds	May–June	Outdoors
PICEA	Spruce	C	Seeds	March–April	Outdoors
			Grafting (veneer)	March	Greenhouse
			Semi-ripe cuttings	Sept–Oct	Frame
PIERIS		S T	Semi-ripe cuttings	Aug–Sept	Frame
			Seeds	March–April	Frame
			Simple or air layering	April–Aug	Outdoors

Genus	Common name	Code	Method	Time	Location
PILEOSTEGIA		Cl	Simple or serpentine layering	April–Aug	Outdoors
			Semi-ripe cuttings	July–Aug	Greenhouse
PIMPINELLA	Aniseed	An	Seeds	March–April	Outdoors
PINGUICULA	Butterwort	P	Division	March–April	Outdoors
			Seeds	March–April	Greenhouse
PINUS	Pine	C	Seeds	March–April	Outdoors
			Grafting (veneer)	March	Greenhouse
PIPTANTHUS		S	Seeds	March–April	Greenhouse
PITTOSPORUM		T S	Seeds	March–April	Greenhouse
			Semi-ripe cuttings	July–Aug	Greenhouse
PLANTAGO	Plantain	P	Seeds	May–June	Outdoors
			Division	March–April	Outdoors
PLATANUS	Plane	T	Seeds	March–April	Outdoors
			Hardwood cuttings	Nov–Dec	Frame
			Air layering	April–Aug	Outdoors
PLATYCODON	Chinese or Japanese Bellflower	P	Basal cuttings	Spring	Greenhouse
			Seeds	May–June	Outdoors
PLATYSTEMON		An	Seeds	March–April or Sept	Outdoors
POA		G (P An)	Seeds	May–June	Outdoors
			Division	March–April	Outdoors

Plant	Type	Method	Time	Place
PODOCARPUS	C	Seeds	March–April	Greenhouse
		Semi-ripe cuttings	July–Aug	Greenhouse/frame
PODOPHYLLUM	P	Division	March–April	Outdoors
		Seeds	May–June	Outdoors
POLEMONIUM Jacob's Ladder	P	Division	March–April	Outdoors
		Seeds	May–June	Outdoors
POLYGALA Milk Wort	A	Soft or semi-ripe cuttings	June–July	Greenhouse/frame
		Division (where possible)	March–April	Outdoors
POLYGONATUM Solomon's Seal	P	Division	March–April	Outdoors
		Seeds	When ripe	Frame
		Eyes on rhizomes	Spring	Greenhouse/frame
POLYGONUM trailing species	P	Division	March–April	Outdoors
		Semi-ripe cuttings	Aug	Frame
baldschuanicum	Cl	Hardwood cuttings	Dec	Greenhouse
POLYPODIUM Polypody	F	Division	March–April	Outdoors
		Spores	When ripe	Greenhouse
POLYSTICHUM Shield and Holly Ferns	F	Division	March–April	Outdoors
		Spores	When ripe	Greenhouse
		Buds on leaves	Summer	Greenhouse
PONCIRUS Japanese Bitter Orange	S	Seeds	When ripe	Frame
		Semi-ripe cuttings	Aug–Sept	Greenhouse

		Aq		April–June	Outdoors
PONTEDERIA	Pickerel Weed	Aq	Division	April–June	Outdoors
POPULUS	Poplar	T	Hardwood cuttings	Nov–Dec	Outdoors
			Suckers	March–April	Outdoors
POTAMOGETON	Pondweed	Aq	Soft cuttings	April–June	Outdoors
			Division	April–June	Outdoors
POTENTILLA	Cinquefoil	S	Semi-ripe cuttings	Sept–Oct	Frame/low polythene tunnel
			Seeds	March–April	Greenhouse
		P A	Division	March–April	Outdoors
			Seeds	Jan–Feb or	Outdoors
				May–June	
PRATIA		A	Division	March–April	Outdoors
			Semi-ripe cuttings	June–July	Greenhouse
			Seeds	Jan–Feb	Frame
			Layering	Spring	Outdoors
PRIMULA	including Primroses, Cowslips, Auriculas	P A	Seeds	When ripe	Frame/greenhouse
			Division	March–April or after flowering if spring flowering	Outdoors
denticulata	Drumstick Primrose	P	Root cuttings	Dec–Jan	Frame/greenhouse
			Division	After flowering	Outdoors
x variabilis	Polyanthus	P	Seeds	Jan–March	Greenhouse
				April	Frame
				Aug	Frame
PRUNELLA	Self Heal	P	Division	March–April	Outdoors

Plant	Type	Method	Time	Place
PRUNUS				
Almonds, Apricots, Cherries (including ornamental), Damsons, Nectarines, Peaches, Plums	T	Budding Grafting (whip-and-tongue)	July–Aug March	Outdoors Outdoors
glandulosa	S	Softwood cuttings	May–June	Greenhouse
laurocerasus Cherry Laurel	S	Semi-ripe cuttings	Sept–Oct	Frame/low polythene tunnel/outdoors
lusitanica Portugal Laurel	S	Semi-ripe cuttings	Sept–Oct	Frame/low polythene tunnel
pumila Dwarf Cherry	S	Softwood cuttings	May–June	Greenhouse
all species of prunus	T S	Seeds	March–April	Outdoors
PSEUDOLARIX Golden Larch	C	Seeds	March–April	Greenhouse/frame
PSEUDOTSUGA Douglas Fir	C	Seeds	March–April	Outdoors
PTELEA Hop Tree	S T	Seeds	When ripe or March–April	Frame
		Simple layering	April–Aug	Outdoors
PTEROCARYA Wingnut	T	Seeds	When ripe or March–April	Frame
		Rooted suckers	March	Outdoors
		Air or simple layering	April–Aug	Outdoors
PULMONARIA Lungwort	P	Division	March–April or after flowering	Outdoors
		Seeds	May–June	Outdoors

Note: Plant entries — "glandulosa Dwarf Cherry", "pumila Dwarf Cherry"

PULSATILLA vulgaris	Pasque Flower	P	Seeds	When ripe	Outdoors
		P	Root cuttings	Dec–Jan	Greenhouse
PUSCHKINIA	Squill	B	Bulblets	Sept	Outdoors
			Seeds	When ripe or Jan–Feb	Outdoors
PYRACANTHA	Fire Thorn	S	Semi-ripe cuttings	Aug	Greenhouse/frame
			Seeds	March–April	Outdoors
PYRETHRUM		P	Basal cuttings	Spring	Greenhouse
			Division	March–April or after flowering	Outdoors
			Seeds	May–June	Outdoors
PYRUS	Pear (fruiting and ornamental)	T	Seeds	March–April	Outdoors
			Budding	July	Outdoors
			Grafting (whip-and-tongue)	March	Outdoors
QUERCUS	Oak	T	Seeds	When ripe	Outdoors
			Grafting (spliced side veneer)	March	Greenhouse
RAMONDA	Rosette Mullein	A	Leaf cuttings	June–July	Frame/greenhouse
			Division	March–April	Outdoors
			Seeds	Jan–Feb	Greenhouse/frame

Plant	Type	Method	Time	Place
RANUNCULUS Buttercup, Crowfoot	P A	Seeds	When ripe	Outdoors
		Division	March–April	Outdoors
aquatilis Water Buttercup, Water Crowfoot	Aq	Division	April–June	Outdoors
lingua Greater Spearwort	Aq	Division	April–June	Outdoors
RAOULIA	A	Division	March–April	Outdoors
		Seeds	Jan–Feb	Outdoors
RESEDA Mignonette	An	Seeds	March–April	Outdoors
REYNOUTRIA (*Polygonum*)	P	Division	March–April	Outdoors
RHAMNUS Buckthorn	S	Seeds	March–April	Outdoors
		Semi-ripe cuttings	July–Aug	Greenhouse
RHAPHIOLEPIS	S	Semi-ripe cuttings	July–Aug	Greenhouse
RHAZYA	P	Division	March–April	Outdoors
		Seeds	May–June	Outdoors
RHEUM Rhubarb (including ornamental)	P	Division	March–April	Outdoors
		Seeds	May–June	Outdoors
RHODODENDRON species	S T	Seeds	March–April	Greenhouse
		Simple layering	April–Aug	Outdoors
evergreens, e.g. hardy hybrids	S	Semi-ripe cuttings	Sept–Oct	Greenhouse
		Grafting (saddle, spliced side)	March	Greenhouse
		Simple layering	April–Aug	Outdoors

evergreen azaleas	S	Semi-ripe cuttings	June–Aug	Greenhouse
		Simple layering	April–Aug	Outdoors
deciduous species and hybrids	S	Softwood cuttings	April–June	Greenhouse
		Simple layering	April–Aug	Outdoors
deciduous azaleas	S	Softwood cuttings	April–June	Greenhouse
		Simple layering	April–Aug	Outdoors
RHODOHYPOXIS	A	Division	April	Outdoors
		Seeds	Jan–Feb	Greenhouse/frame
RHODOTYPOS White Kerria	S	Semi-ripe cuttings	June–July	Greenhouse
		Hardwood cuttings	Oct–Nov	Frame
RHUS Sumach	S T	Root cuttings	Dec–Jan	Frame
		Rooted suckers	March–April	Outdoors
		Seeds	March–April	Frame/greenhouse
RIBES nigrum and sativum Flowering Currant Black, Red and White Currants	S	Hardwood cuttings	Nov–Dec	Frame/outdoors
	S	Hardwood cuttings	Nov–Dec	Outdoors/frame
grossularia Gooseberry	S	Hardwood cuttings	Nov–Dec	Outdoors/frame
ROBINIA False Acacia, Locust	T S	Seeds	March–April	Greenhouse/frame
		Root cuttings	Dec–Jan	Frame
		Suckers	March–April	Outdoors
		Grafting (whip-and-tongue)	March	Outdoors
RODGERSIA	P	Division	March–April	Outdoors
		Seeds	March–April	Frame
ROMNEYA Californian Tree Poppy	P	Root cuttings	Dec–Jan	Greenhouse

Plant	Type	Method	Time	Place
ROSA species Rose	S C1	Seeds (inclined to hybridize)	March–April	Outdoors
hybrids **Hybrid Teas, Floribundas, Shrub, Climbers, Ramblers, Miniatures**		Budding	June–Sept	Outdoors
Ramblers, and other strong growers		Hardwood cuttings	Nov–Dec	Outdoors
ROSCOEA	P	Seeds	When ripe	Frame
		Division	Spring – when shoots are visible	Outdoors
ROSMARINUS Rosemary	S	Semi-ripe cuttings	Sept–Oct	Frame/low polythene tunnel
RUBUS Blackberry	C1	Tip layering	July–Aug	Outdoors
		Leaf-bud cuttings	Aug	Frame
Loganberry	C1	Tip layering	July–Aug	Outdoors
		Leaf-bud cuttings	Aug	Frame
Raspberry	S	Rooted suckers	Nov–Dec	Outdoors
odoratus **Ornamental Brambles**	S	Division	March–April	Outdoors
spectabilis parviflorus				
cockburnianus **Ornamental Bramble**	S	Root cuttings	Dec–Jan	Frame/greenhouse
many species **Ornamental Brambles**	S	Simple layering	Summer	Outdoors
RUDBECKIA Coneflower	P	Division	March–April	Outdoors

		An	Seeds	May–June	Outdoors
RUSCUS	Butcher's Broom	S	Division	Feb–March	Greenhouse
			Seeds	March	Greenhouse
RUTA	Rue	S	Seeds	March–April	Outdoors
				March–April	Frame
			Semi-ripe cuttings	March–April	Frame/outdoors
				Sept–Oct	Frame/low polythene tunnel
SAGINA subulata 'Aurea'	Heath Pearlwort	P	Division	Autumn/ spring	Outdoors/frame
SAGITTARIA	Arrowhead	Aq	Division	April–June	Outdoors
SALIX	Willow, Sallow	T S	Hardwood cuttings	Nov–Dec	Outdoors
SALVIA	Sage	S	Semi-ripe cuttings	June–July	Frame
			Seeds	March–April	Greenhouse
		P	Division	March–April	Outdoors
			Seeds	March–April	Greenhouse
horminum	Clary	An	Seeds	March–April	Outdoors
SAMBUCUS	Elder	S	Semi-ripe cuttings	July–Aug	Greenhouse
			Hardwood cuttings	Nov–Dec	Frame/outdoors
			Seeds	March–April	Outdoors
SANGUINARIA	Bloodroot	P	Division	Oct–Nov	Outdoors
SANGUISORBA	Burnet	P	Division	March–April	Outdoors
			Seeds	April	Frame
SANTOLINA	Lavender Cotton	S	Semi-ripe cuttings	Sept–Oct	Frame/low polythene tunnel
			Hardwood cuttings	Nov–Dec	Outdoors

Plant		Type	Method	Time	Place
SANVITALIA	Creeping Zinnia	An	Seeds	March–April	Outdoors
SAPONARIA	Soapwort	A	Softwood cuttings	June	Greenhouse/frame
			Seeds	Jan–Feb	Outdoors
		An	Seeds	March–April	Outdoors
SARCOCOCCA		S	Division	March–April	Outdoors
			Semi-ripe cuttings	Oct	Frame
			Seeds	March–April	Outdoors
SASSAFRAS		T	Seeds	Feb–March	Greenhouse
			Rooted suckers	Autumn–spring	Outdoors
			Root cuttings	Dec–Jan	Frame/greenhouse
SAXIFRAGA	Saxifrage	A	Seeds	Jan–Feb	Outdoors
			Division	March–April	Outdoors
			Cuttings (single rosettes)	June–July	Greenhouse/frame
SCABIOSA	Scabious	P	Division	March–April	Outdoors
			Seeds	May–June	Outdoors
		An	Seeds	March–April	Outdoors
				Sept	Outdoors
SCHIZANDRA		Cl	Semi-ripe cuttings	Aug	Greenhouse
			Seeds	When ripe	Frame
SCHIZOPHRAGMA		Cl	Seeds	March–April	Greenhouse
			Semi-ripe cuttings	Aug	Greenhouse

			Simple or serpentine layering	April—Aug	Outdoors
SCHIZOSTYLIS		P	Division	April	Outdoors
			Seeds	April—May	Greenhouse
SCIADOPITYS	Umbrella Pine	C	Seeds	March—April	Greenhouse
SCILLA	Squill	B	Seeds	When ripe or Jan—Feb	Outdoors
			Bulblets	Sept	Outdoors
SCIRPUS	Clubrush	Aq	Division	April—June	Outdoors
SCROPHULARIA		P	Division	March—April	Outdoors
			Basal cuttings	May—June	Greenhouse/frame
SCUTELLARIA	Skull Cap	P	Division	March—April	Outdoors
			Seeds	April	Frame
SEDUM	Stonecrop	P A	Seeds	March—April	Frame/greenhouse
			Soft cuttings	June—July	Frame/greenhouse
			Leaf cuttings	June—July	Frame/greenhouse
			Division	March—April	Outdoors
SEMPERVIVUM	Houseleek	A	Division	March—April	Outdoors
			Rooted offsets	Summer	Outdoors
SENECIO		S	Semi-ripe cuttings	Sept—Oct	Frame/low polythene tunnel
		P	Division	March—April	Outdoors
		S P	Seeds	March—April	Greenhouse/frame
SEQUOIA	Redwood	C	Seeds	March—April	Outdoors
			Semi-ripe cuttings	Sept—Oct	Frame

255

Plant		Type	Method	Time	Place
SEQUOIADENDRON	Wellingtonia	C	Seeds	March–April	Outdoors
SHORTIA		P	Division	March–April	Outdoors
			Seeds	April	Frame
SIDALCEA	Greek Mallow	P	Division	March–April	Outdoors
			Seeds	May–June	Outdoors
			Basal cuttings	Spring	Greenhouse
SILENE	Catchfly, Campion	P A	Division	March–April	Outdoors
			Soft basal cuttings	April–May	Frame/greenhouse
			Seeds	May–June	Outdoors
		An	Seeds	March–April	Outdoors
SILYBUM	Milk Thistle	Bi	Seeds	May–June	Outdoors
SISYRINCHIUM	Satin Flower	P	Division	March–April	Outdoors
			Seeds	May–June	Outdoors
SKIMMIA		S	Semi-ripe cuttings	Sept–Oct	Frame/low polythene tunnel
			Simple layering	April–Aug	Outdoors
			Seeds	March–April	Frame
SMILACINA		P	Division	March–April	Outdoors
SOLANUM		S Cl	Semi-ripe cuttings	July–Aug	Greenhouse
SOLDANELLA		P A	Division	March–April	Outdoors
			Seeds	When ripe	Outdoors/frame
SOLIDAGO	Golden Rod	P	Division	March–April	Outdoors
			Seeds	April	Frame

	P	Division	March—April	Outdoors
x SOLIDASTER	P	Division	March—April	Outdoors
SOPHORA	T S	Seeds	March—April	Greenhouse
SORBARIA	S	Rooted suckers	Autumn—spring	Outdoors
		Hardwood cuttings	Nov—Dec	Frame
		Root cuttings	Dec—Jan	Frame
SORBUS including Mountain Ash and Whitebeam	T	Seeds	March—April	Outdoors
		Grafting (whip-and-tongue)	March	Outdoors
		Budding	July	Outdoors
SPARAXIS	Cm	Cormlets	Oct—Nov	Outdoors
		Seeds	Autumn	Outdoors
SPARTIUM Spanish Broom	S	Seeds	March—April	Greenhouse
SPIRAEA	S	Softwood cuttings	June—July	Greenhouse/frame
		Hardwood cuttings	Nov—Dec	Frame
		Division (where possible)	March—April	Outdoors
STACHYS Woundwort	P	Division	March—April	Outdoors
		Seeds	April	Frame
STACHYURUS	S	Simple layering	April—Aug	Outdoors
		Semi-ripe cuttings	July—Aug	Greenhouse
STAPHYLEA Bladder Nut	S T	Seeds	March—April	Greenhouse
		Simple layering	April—Aug	Outdoors
		Semi-ripe cuttings	July—Aug	Greenhouse
		Rooted suckers	March	Outdoors

Plant	Type	Method	Time	Place
STEPHANANDRA	S	Division	March–April	Outdoors
		Semi-ripe cuttings	July	Frame
		Hardwood cuttings	Nov	Outdoors
STERNBERGIA	B	Bulblets	Aug–Sept	Outdoors
		Seeds	When ripe or Jan–Feb	Outdoors
STEWARTIA	S T	Seeds	March–April	Greenhouse
		Softwood cuttings	June	Greenhouse
		Simple or air layering	April–Aug	Outdoors
STIPA	G (P)	Division	March–April	Outdoors
		Seeds	May–June	Outdoors
STOKESIA Stokes' Aster	P	Division	March–April	Outdoors
		Seeds	May–June	Outdoors
		Root cuttings	Nov–Dec	Frame
STRANVAESIA	S T	Seeds	March–April	Outdoors
		Simple layering	April–Aug	Outdoors
STRATIOTES Water Soldier	Aq	Offsets (when large enough)	Spring–summer	Outdoors
STYRAX	T S	Seeds	March–April	Greenhouse
		Simple or air layering	April–Aug	Outdoors
		Semi-ripe cuttings	June–July	Greenhouse

SYCOPSIS		S T	Simple or air layering	April–Aug	Outdoors
			Seeds	March–April	Greenhouse
SYMPHORICARPOS	Snowberry	S	Division (rooted suckers)	March–April	Outdoors
			Hardwood cuttings	Nov–Dec	Frame/outdoors
			Seeds	March–April	Outdoors
SYMPHYANDRA	Ring Bellflower	P	Division	March–April	Outdoors
			Seeds	May–June	Outdoors
			Soft cuttings	April–June	Greenhouse
SYMPHYTUM	Comfrey	P	Division	March–April	Outdoors
			Seeds	May–June	Outdoors
			Root cuttings	Dec–Jan	Frame
SYMPLOCOS		S T	Simple or air layering	April–Aug	Outdoors
			Semi-ripe cuttings	July	Greenhouse
			Seeds	March–April	Frame
SYRINGA	Lilac	S	Simple layering	April–Aug	Outdoors
			Semi-ripe cuttings	Aug–Sept	Greenhouse
			Seeds	March–April	Frame
			Grafting (spliced side veneer)	March	Greenhouse
TAMARIX	Tamarisk	S T	Hardwood cuttings	Nov–Dec	Outdoors/frame

Plant	Type	Method	Time	Place
TANACETUM	P	Division	March–April	Outdoors
		Seeds	May–June	Outdoors
		Soft basal cuttings	April–June	Outdoors
TAXODIUM Swamp Cypress	C	Seeds	March–April	Frame/outdoors
		Hardwood cuttings	Nov–Dec	Greenhouse
TAXUS Yew	C	Seeds	March–April	Outdoors
		Semi-ripe cuttings	Aug–Oct	Greenhouse/frame
		Grafting (veneer)	March	Greenhouse
TECHOPHILAEA Chilean Crocus	Cm	Cormlets	Sept or spring	Outdoors
		Seeds	March–April	Greenhouse
TELLIMA	P	Division	March–April	Outdoors
		Seeds	May–June	Outdoors
TEUCRIUM Germander	S P	Semi-ripe cuttings	June–Aug	Frame
		Division	March–April	Outdoors
		Seeds	March–April	Greenhouse/frame
THALICTRUM Meadow Rue	P	Seeds	May–June	Outdoors
		Division	March–April	Outdoors
		Soft basal cuttings	April–May	Frame
'Hewitt's Double' (double form)		Division	March–April	Outdoors
		Soft basal cuttings	April–May	Frame
THELYPTERIS	F	Division	March–April	Outdoors
		Spores	When ripe	Greenhouse

Genus	Common name	Type	Method	Time	Location
THERMOPSIS		P	Division Seeds	March–April March–April	Outdoors Greenhouse
THLASPI	Penny Cress	P	Seeds Division Soft cuttings	March–April March–April April–May	Outdoors Outdoors Greenhouse/frame
THUJA	Arbor-vitae	C	Seeds Semi-ripe cuttings	March–April Sept–Oct	Outdoors Frame/greenhouse
THUJOPSIS	Hiba Arbor-vitae	C	Seeds Semi-ripe cuttings	March–April Sept–Oct	Outdoors Frame/greenhouse
THYMUS	Thyme	S A	Division Seeds Softwood cuttings	March–April Jan–Feb June	Outdoors Outdoors Greenhouse/frame
TIARELLA	Foam Flower	P	Division Seeds	March–April May–June	Outdoors Outdoors
TIGRIDIA	Tiger Flower	B	Bulblets Seeds	April Jan–Feb	Outdoors Greenhouse
TILIA	Lime, Linden	T	Seeds Simple or air layering	When ripe April–June	Outdoors Outdoors
TOLMIEA	Pick-a-back Plant	P	Division Plantlets on leaves	March–April Summer	Outdoors Frame/greenhouse
TORREYA		C	Seeds Semi-ripe cuttings	March–April Aug–Sept	Greenhouse Frame/greenhouse

Plant		Type	Method	Time	Place
TRACHELIUM	Throatwort	P	Seeds	March–April	Greenhouse
			Soft cuttings	May–June	Frame/greenhouse
TRACHELOSPERMUM		Cl	Simple or serpentine layering	April–Aug	Outdoors
			Semi-ripe cuttings	Aug	Greenhouse
TRACHYCARPUS	Chusan Palm	T	Seeds	March–April	Greenhouse
TRACHYSTEMON		P	Seeds	May–June	Outdoors
			Division	March–April	Outdoors
TRADESCANTIA	Spiderwort	P	Division	March–April	Outdoors
			Seeds	March–April	Frame
TRAGOPOGON		P B	Seeds	May–June	Outdoors
TRICYRTIS	Toad Lily	P	Division	March–April	Outdoors
			Seeds	March–April	Frame
TRIFOLIUM	Clover, Trefoil	P	Division	March–April	Outdoors
TRILLIUM	Wood Lily	P	Division	March–April	Outdoors
			Seeds	When ripe	Frame
TRITONIA		Cm	Cormlets	Oct–Nov	Outdoors
			Seeds	Autumn	Outdoors
TROCHODENDRON		T	Air or simple layering	April–Aug	Outdoors

TROLLIUS	Globe Flower	P	Division	March–April	Outdoors
			Seeds	When ripe	Frame
TROPAEOLUM	Nasturtium	An	Seeds	April	Outdoors
		P(Cl)	Seeds	March–April	Greenhouse/frame
			Soft cuttings	April–May	Greenhouse
TSUGA	Hemlock	C	Seeds	March–April	Outdoors
			Semi-ripe cuttings	July–Sept	Frame
TULIPA	Tulip	B	Bulblets	Oct–Nov	Outdoors
			Seeds	When ripe	Frame
TYPHA	Reedmace	Aq	Division	April–June	Outdoors
ULEX	Gorse	S	Seeds	March–April	Greenhouse
			Semi-ripe cuttings	Aug	Greenhouse/frame
ULMUS	Elm	T	Seeds	When ripe	Outdoors
			Rooted suckers	March	Outdoors
			Simple or air layering	April–Aug	Outdoors
			Softwood cuttings	June	Greenhouse
			Hardwood cuttings	Nov–Dec	Greenhouse/frame
UMBELLULARIA	California Laurel	T	Seeds	March–April	Greenhouse
			Semi-ripe cuttings	July–Aug	Greenhouse/frame
			Air or simple layering	April–Aug	Outdoors
UTRICULARIA	Bladderwort	Aq	Division	April–June	Outdoors
UVULARIA	Bellwort	P	Division	March–April	Outdoors

Plant		Type	Method	Time	Place
VACCINIUM	**Blueberry, Bilberry**	**S**	Seeds	March–April	Greenhouse
			Layering (dropping)	April	Outdoors
			Semi-ripe cuttings	Aug	Greenhouse/frame
VALERIANA	**Valerian**	**P**	Division	March–April	Outdoors
			Soft cuttings	April–May	Greenhouse/frame
			Seeds	April–May	Outdoors
VANCOUVERIA		**P**	Division	March–April	Outdoors
VERATRUM	**False Hellebore**	**P**	Seeds	March–April	Greenhouse/frame
			Division	March	Outdoors
VERBASCUM	**Mullein**	**P**	Seeds	May–June	Outdoors
			Division	March–April	Outdoors
			Root cuttings	Dec–Jan	Greenhouse/frame
		Bi	Seeds	May–June	Outdoors
VERBENA	**Vervain**	**P**	Seeds	Jan–March	Greenhouse
			Division	March–April	Outdoors
			Semi-ripe cuttings	Aug	Greenhouse/frame
VERNONIA		**P**	Division	March–April	Outdoors
VERONICA	**Speedwell**	**P A**	Division	March–April	Outdoors
			Seeds	May–June	Outdoors
			Soft cuttings	May–June	Greenhouse/frame

Plant			Method	Time	Location
VIBURNUM		S	Seeds	March–April	Greenhouse/frame
			Semi-ripe cuttings	July–Sept	Greenhouse/frame
			Hardwood cuttings	Nov–Dec	Frame
			Simple layering	April–Aug	Outdoors
VINCA	Periwinkle	S	Semi-ripe cuttings	Aug–Oct	Frame/low polythene tunnel
			Division	March–April	Outdoors
			Serpentine layering	April–Aug	Outdoors
VIOLA	Pansy, Violet	P	Seeds	March–April	Frame/greenhouse
			Division	March–April	Outdoors
winter/spring flowering pansies			Cuttings	Aug	Frame
summer-flowering pansies			Seeds	June–July	Frame
			Seeds	Feb–March	Greenhouse/frame
VISCARIA		A	Seeds	Jan–Feb	Outdoors
			Division	March–April	Outdoors
VITIS	Vines, including grape vines	Cl	Eye cuttings	Dec–Jan	Greenhouse
			Seeds	March–April	Greenhouse
			Serpentine layering	April–Aug	Outdoors
WAHLENBERGIA		P A	Division	March–April	Outdoors
			Seeds	Jan–Feb	Outdoors
			Softwood cuttings	April–June	Frame
WALDSTEINIA		P	Division	March–April	Outdoors

Plant		Type	Method	Time	Place
WEIGELA		S	Semi-ripe cuttings	June–Aug	Greenhouse
			Hardwood cuttings	Nov–Dec	Frame
			Seeds	March–April	Outdoors
WISTERIA		Cl	Simple or serpentine layering	April–Aug	Outdoors
			Hardwood cuttings	Nov–Dec	Greenhouse
			Seeds	March–April	Greenhouse
WOODWARDIA virginiana	Chain Fern	F	Division	March	Frame
XERANTHEMUM	Immortelle	An	Seeds	March–April	Outdoors
YUCCA		S T	Seeds	March–April	Greenhouse
			'Toes', buds situated on stem bases or rhizomes. Remove toes and plant.	April–May	Frame/outdoors
			Rooted suckers	March–April	Outdoors
ZANTEDESCHIA aethiopica	Arum Lily	P	Seeds	March–April	Greenhouse
			Offsets (division)	Spring	Outdoors
			Bulbils on roots	Spring	Frame/greenhouse

ZANTHOXYLUM	Prickly Ash	T	Seeds	March–April	Greenhouse
			Root cuttings	Dec–Jan	Greenhouse/frame
ZAUSCHNERIA	California Fuchsia	P	Semi-ripe cuttings	Aug–Sept	Greenhouse/frame
			Division	March–April	Outdoors
			Seeds	March	Greenhouse
ZELKOVA		T	Seeds	March–April	Outdoors/frame
			Air or simple layering	April–Aug	Outdoors
ZENOBIA		S	Seeds	March–April	Frame
			Simple layering	April–Aug	Outdoors
ZEPHYRANTHES	Zephyr Lily	B	Bulblets	April	Outdoors
			Division of clumps	April	Outdoors
			Seeds	Jan–Feb	Greenhouse/frame

Plant		Type	Method	Time	Place
ABUTILON		**S**	Seeds	Jan–April	Greenhouse
			Semi-ripe cuttings	July–Sept	Greenhouse/indoors
ACACIA	Wattle	**S T**	Seeds	When ripe	Greenhouse
			Semi-ripe cuttings	July–Aug	Greenhouse
ACALYPHA		**S**	Softwood cuttings	April	Greenhouse
ACHIMENES		**P**	Seeds	Feb–March	Greenhouse
			Softwood cuttings	Feb–March	Greenhouse/indoors
			Leaf cuttings	Feb–March	Greenhouse/indoors
ACIDANTHERA		**Cm**	Seeds	Feb–March	Greenhouse
			Cormlets	April	Greenhouse
ADIANTUM	Maidenhair Fern	**F**	Division	March–April	Greenhouse/indoors
			Spores	When ripe	Greenhouse
ANDROMISCHUS		**Suc**	Leaf cuttings	June–July	Greenhouse/indoors
AECHMEA		**P**	Offsets	April–July	Greenhouse/indoors
			Seeds	March–April	Greenhouse
AEONIUM		**Suc**	Seeds	April	Greenhouse/indoors
			Leaf or stem cuttings	April–July	Greenhouse/indoors
AERIDES		**Orch**	Cuttings	April–May	Greenhouse
AESCHYNANTHUS		**S**	Softwood cuttings	April–May	Greenhouse
AGAPETES		**S**	Seeds	April	Greenhouse
			Semi-ripe cuttings	Aug–Sept	Greenhouse

AGAVE	Century Plant	P	Offsets Seeds	April Jan–April	Greenhouse/indoors Greenhouse/indoors
AGERATUM	Floss Flower	An	Seeds	Feb–March	Greenhouse
AGLAONEMA		P	Division Semi-ripe cuttings Cuttings (stem sections)	April July–Aug July–Aug	Greenhouse/indoors Greenhouse Greenhouse
AICHRYSON		Suc	Seeds Cuttings	April April–July	Greenhouse/indoors Greenhouse/indoors
ALLAMANDA		Cl	Softwood cuttings	April	Greenhouse
ALOE		Suc	Offsets Seeds Leaf cuttings	April Jan–April April–Aug	Greenhouse/indoors Greenhouse/indoors Greenhouse/indoors
ALONSOA		P S	Seeds	March	Greenhouse
ALTERNANTHERA		P	Division Semi-ripe cuttings	April Aug	Greenhouse Greenhouse
AMARANTHUS		An	Seeds	April	Greenhouse/outdoors
AMMOBIUM	Sand Immortelle	An	Seeds	March	Greenhouse
ANANAS	Pineapple	P	Suckers or offsets	April	Greenhouse/indoors
ANGRAECUM		Orch	Side shoots, as cuttings	May–June	Greenhouse
ANIGOZANTHUS	Kangaroo Paw	P	Division Seeds	April–May Jan–April	Greenhouse Greenhouse

Plant	Type	Method	Time	Place
ANTHOLYZA	Cm	Cormlets	Feb–March	Greenhouse
		Seeds	Sept	Greenhouse
ANTHURIUM	P	Division	Jan	Greenhouse/indoors
		Seeds	When ripe	Greenhouse
ANTIRRHINUM Snapdragon	An	Seeds	Jan–Feb	Greenhouse
APHELANDRA	S	Softwood cuttings	April–July	Greenhouse
		Seeds	Jan–April	Greenhouse
		Leaf-bud cuttings	April–July	Greenhouse
APOROCACTUS Rat-tail Cactus	Ca	Cuttings	June–Aug	Greenhouse/indoors
		Seeds	Jan–April	Greenhouse/indoors
ARAUCARIA excelsa Norfolk Island Pine	C	Seeds	Jan–April	Greenhouse
ARCTOTIS African Daisy	P	Semi-ripe cuttings	Aug–Sept	Greenhouse
	An	Seeds	March–April	Greenhouse
ARDISIA	T S	Soft or semi-ripe cuttings (cut back plants to encourage shoots)	March–Sept	Greenhouse
		Seeds	April	Greenhouse
ASCLEPIAS Milkweed	P	Seeds	Jan–April	Greenhouse
		Softwood cuttings	May–July	Greenhouse
ASPARAGUS	P	Division	April	Greenhouse/indoors
		Seeds	Jan–April	Greenhouse/indoors

		P	Division	April	Greenhouse/indoors
ASPIDISTRA	Cast-iron Plant	P	Division	April	Greenhouse/indoors
ASPLENIUM	Spleenwort	F	Division	April	Greenhouse/indoors
			Plantlets on leaves	When available	Greenhouse/indoors
			Spores	As soon as ripe	Greenhouse
ASTROPHYTUM	Star Cactus	Ca	Seeds	Jan–April	Greenhouse/indoors
ASYSTASIA		S	Softwood cuttings	April–June	Greenhouse
BABIANA		Cm	Seeds	April	Greenhouse
			Cormlets	Oct	Greenhouse
BANKSIA		S T	Seeds	As soon as ripe	Greenhouse
			Semi-ripe cuttings	July–Oct	Greenhouse
BAROSMA		S	Semi-ripe cuttings	Aug–Sept	Greenhouse
			Seeds	Jan–April	Greenhouse
BEGONIA		P	Seeds	Jan–Feb	Greenhouse
			Soft cuttings	April–June	Greenhouse/indoors
			Division	April	Greenhouse/indoors
masoniana			Leaf cuttings	April–June	Greenhouse/indoors
rex		Tu	Seeds	Jan–Feb	Greenhouse
			Division	April	Greenhouse
BELOPERONE	Shrimp Plant	S	Soft or semi-ripe cuttings	June–Aug	Greenhouse/indoors

Plant	Type	Method	Time	Place
BILLARDIERA	Cl	Seeds	Jan–April	Greenhouse
		Semi-ripe cuttings	Aug–Sept	Greenhouse
BILLBERGIA	P	Division	April	Greenhouse/indoors
		Seeds	March–April	Greenhouse
BLECHNUM	F	Division	April	Greenhouse/indoors
		Spores	When ripe	Greenhouse
BLETIA	Orch	Division	After flowering	Greenhouse
BLETILLA	Orch	Division	After flowering	Greenhouse
BOMAREA	Cl	Seeds	April	Greenhouse
		Division	April	Greenhouse
BORONIA	S	Seeds	Jan–April	Greenhouse
		Semi-ripe cuttings	Aug–Sept	Greenhouse
BORZICACTUS	Ca	Seeds	April	Greenhouse/indoors
BOUGAINVILLEA	S	Semi-ripe cuttings	July–Aug	Greenhouse
BOUVARDIA	S	Softwood cuttings	April–June	Greenhouse
BRASSIA	Orch	Division	April	Greenhouse
x BRASSOCATTLEYA	Orch	Division	April	Greenhouse
BROWALLIA	An	Seeds	March–April	Greenhouse
			Sept	Greenhouse

Name	Common name		Method	Timing	Location
BRUNFELSIA		S	Semi-ripe cuttings	June–Aug	Greenhouse
BRYOPHYLLUM (also known as *Kalanchoe*)		Suc	Plantlets on leaves	When formed	Greenhouse/indoors
			Cuttings	April–June	Greenhouse/indoors
			Seeds	Jan–April	Greenhouse/indoors
CAESALPINIA		T S	Seeds	Jan–April	Greenhouse
		Cl	Simple layering	April–June	Greenhouse
CALADIUM		Tu	Divide dormant tubers	Feb–March	Greenhouse
CALANDRINIA	Rock Purslane	An	Seeds	March	Greenhouse
CALANTHE		Orch	Division	When potting	Greenhouse
CALATHEA		P	Division	April	Greenhouse/indoors
CALCEOLARIA	Slipper Wort	An	Seeds	Feb–March June	Greenhouse
		S	Softwood or semi-ripe cuttings	April or Aug–Sept	Greenhouse/frame
CALLISIA		P	Softwood or semi-ripe cuttings	April–Sept	Greenhouse
CALLISTEMON	Bottlebrush	S	Seeds	Jan–April	Greenhouse
			Semi-ripe cuttings	June–Aug	Greenhouse
CALLISTEPHUS	China Aster	An	Seeds	March–April	Greenhouse
CAMPANULA isophylla	Bell Flower	P	Seeds	March	Greenhouse
			Soft cuttings	April–June	Greenhouse/indoors

Plant		Type	Method	Time	Place
CANNA	Indian Shot Lily	P	Division	April	Greenhouse
			Seeds	Feb	Greenhouse
CAPSICUM	Ornamental Peppers	An	Seeds	April	Greenhouse
CARNEGIEA		Ca	Seeds	Jan–April	Greenhouse/indoors
CASSIA		T S	Seeds	Jan–April	Greenhouse
			Semi-ripe cuttings	Aug–Sept	Greenhouse
CATTLEYA		Orch	Division	When potting	Greenhouse
CELOSIA	Cockscomb	An	Seeds	Feb–April	Greenhouse
CELSIA		P	Seeds	April	Greenhouse
				June–July	Frame
CENTAUREA		P	Seeds	Feb–March	Greenhouse
			Soft cuttings	March–Sept	Greenhouse
CEPHALOCEREUS	Old Man Cactus	Ca	Seeds	Jan–April	Greenhouse/indoors
CEREUS		Ca	Seeds	Jan–April	Greenhouse/indoors
			Cuttings	June–Aug	Greenhouse/indoors
CEROPEGIA	Hearts Entangled	Cl	Seeds	Jan–April	Greenhouse
			Cuttings	June–Aug	Greenhouse/indoors
			Root the tubers	When	Greenhouse/indoors
			formed at joints	available	
CESTRUM		S	Semi-ripe cuttings	Aug–Sept	Greenhouse
			Seeds	Jan–April	Greenhouse

Name	Common name	Ca	Method	Time	Location
CHAMAECEREUS	Peanut Cactus	Ca	Cuttings	June–Aug	Greenhouse/indoors
			Seeds	Jan–April	Greenhouse/indoors
CHAMAEDOREA suckering species	Palm	P	Seeds	Jan–April	Greenhouse
			Division (suckers)	April	Greenhouse/indoors
CHAMAEROPS	Palm	P	Seeds	Jan–April	Greenhouse
			Rooted suckers	April	Greenhouse/indoors
CHLIDANTHUS		B	Bulblets	April	Greenhouse
CHLOROPHYTUM	Spider Plant	P	Division	April	Greenhouse/indoors
			Layering plantlets	Spring–summer	Greenhouse/indoors
CHORIZEMA		S	Softwood cuttings	March	Greenhouse
			Seeds	Jan–April	Greenhouse
CHRYSANTHEMUM	Late-flowering Chrysanthemum	P	Soft basal cuttings	Jan–March	Greenhouse
frutescens	Marguerite	S	Soft or semi-ripe cuttings	April or Aug	Greenhouse
	Charm, Korean and Cascade Chrysanthemums	P	Seeds	Jan–March	Greenhouse
parthenium		P (An)	Seeds	Feb–March	Greenhouse
CINERARIA		An	Seeds	April–Aug	Greenhouse/frame
CISSUS		Cl	Soft or semi-ripe cuttings	April–Aug	Greenhouse/indoors
		Suc S	Seeds	Jan–April	Greenhouse
			Cuttings	June–Aug	Greenhouse/indoors

Plant	Type	Method	Time	Place
CITRUS — Orange, Lemon, Lime, Grapefruit, Shaddock, Citron, Calamondin	T S	Seeds Semi-ripe cuttings	Jan–April Aug–Sept	Greenhouse/indoors Greenhouse/indoors
CLEISTOCACTUS	Ca	Cuttings Seeds	June–Aug Jan–April	Greenhouse/indoors Greenhouse/indoors
CLEOME — Spider Flower	An	Seeds	Feb–March	Greenhouse
CLERODENDRON	S Cl	Softwood or semi-ripe cuttings Seeds	April–Aug Jan–April	Greenhouse Greenhouse
CLETHRA	S T	Seeds Simple layering Semi-ripe cuttings	March–April April–Aug Aug–Sept	Greenhouse Outdoors Greenhouse
CLIANTHUS	S	Seeds Semi-ripe cuttings	Jan–April Aug–Sept	Greenhouse Greenhouse
CLIVIA	P	Division Rooted offsets Seeds	After flowering Spring–summer When ripe	Greenhouse Greenhouse Greenhouse
COBAEA — Cup and Saucer Vine	Cl	Seeds	Feb–March	Greenhouse
COCOS — Coconut Palm	T	Seeds	Jan–April	Greenhouse

CODIAEUM	Croton	S	Soft or semi-ripe cuttings	April–Aug	Greenhouse
			Air layering	April–Aug	Greenhouse
COELOGYNE		Orch	Division	Spring	Greenhouse
COFFEA	Coffee	S	Semi-ripe cuttings	July–Aug	Greenhouse
			Seeds	Jan–April	Greenhouse
COLEUS		P An	Soft cuttings	April–Aug	Greenhouse/indoors
			Seeds	Jan–April	Greenhouse/indoors
COLQUHOUNIA		S	Semi-ripe cuttings	July–Aug	Greenhouse
			Seeds	Jan–April	Greenhouse
COLUMNEA		P	Soft or semi-ripe cuttings	April–Aug	Greenhouse
			Seeds	Jan–April	Greenhouse
COMMELINA		P	Seeds	March–April	Greenhouse
			Division	March–April	Greenhouse
			Soft or semi-ripe cuttings	June–Aug	Greenhouse
CONOPHYTUM		Suc	Seeds	Jan–April	Greenhouse/indoors
			Division	July	Greenhouse/indoors
COPIAPOA		Ca	Seeds	Jan–April	Greenhouse/indoors
			Offsets	June–Aug	Greenhouse/indoors
COPROSMA		S T	Semi-ripe cuttings	July–Aug	Greenhouse

277

Plant		Type	Method	Time	Place
CORDYLINE		S T	Cuttings (stem sections)	May–June	Greenhouse
			Seeds	Jan–April	Greenhouse
			'Toes' (tuberous root growths)	Any time	Greenhouse
CORONILLA		S	Semi-ripe cuttings	Aug–Sept	Greenhouse
			Seeds	Jan–April	Greenhouse
CORYPHANTHA		Ca	Seeds	Jan–April	Greenhouse/indoors
			Offsets	June–Aug	Greenhouse/indoors
COSMOS	Cosmea	An	Seeds	Feb–March	Greenhouse
				April–May	Outdoors
atrosanguineus		Tu	Cuttings	March–April	Greenhouse
COTULA		An	Seeds	Feb-March	Greenhouse
				April–May	Outdoors
COTYLEDON		Suc	Seeds	Jan–April	Greenhouse/indoors
			Leaf cuttings	June–Aug	Greenhouse/indoors
			Stem cuttings	June–Aug	Greenhouse/indoors
CRASSULA		Suc	Seeds	Jan–April	Greenhouse/indoors
			Leaf cuttings	June–Aug	Greenhouse/indoors
			Stem cuttings	June–Aug	Greenhouse/indoors
x CRINODONNA		B	Division	April	Greenhouse
			Bulblets	April	Greenhouse

CRINUM	B	Division	April	Greenhouse
		Bulblets	April	Greenhouse
		Seeds	When ripe	Greenhouse
CROSSANDRA	P S	Soft or semi-ripe cuttings	April–Aug	Greenhouse
		Seeds	Jan–April	Greenhouse
CRYPTANTHUS Earth Star	P	Rooted suckers or offsets	June	Greenhouse/indoors
CTENANTHE	P	Division	April	Greenhouse/indoors
CUCURBITA Ornamental Gourds	An	Seeds	March	Greenhouse
CUPHEA	P	Softwood cuttings	March–April	Greenhouse
		Seeds	Jan–Feb	Greenhouse
	S	Softwood or semi-ripe cuttings	April–July	Greenhouse
CYANOTIS	P	Semi-ripe cuttings	June–Aug	Greenhouse/indoors
CYCAS	P	Seeds	Jan–April	Greenhouse
		Rooted suckers	Spring–summer	Greenhouse
CYCLAMEN	Tu	Seeds	Aug–March	Greenhouse
CYMBIDIUM	Orch	Division	Feb–April	Greenhouse
		Backbulbs	Feb–April	Greenhouse
CYPERUS Umbrella Plant	P	Division	April	Greenhouse/indoors
		Seeds	Jan–April	Greenhouse
		Leaf rosettes used as cuttings	April–Aug	Greenhouse/indoors

Plant		Type	Method	Time	Place
CYPHOMANDRA	Tree Tomato	T	Seeds	Jan–April	Greenhouse
CYPRIPEDIUM	Lady's Slipper	Orch	Division	April	Greenhouse
CYRTOMIUM		F	Spores	When ripe	Greenhouse
			Division	April	Greenhouse/indoors
CYTISUS		S	Seeds	Jan–April	Greenhouse
			Semi-ripe cuttings	Aug–Sept	Greenhouse
DAHLIA		Tu	Division	March–April (when planting)	Outdoors
			Soft basal cuttings	Feb–March	Greenhouse
			Seeds	Jan	Greenhouse
DATURA	Angel's Trumpets	S	Soft or semi-ripe cuttings	May–Sept	Greenhouse
		An	Seeds	March–April	Greenhouse
DENDROBIUM		Orch	Division	After flowering	Greenhouse
			Plantlets	When available	Greenhouse
DIANTHUS	Perpetual Carnation	P	Softwood cuttings	March	Greenhouse
DICKSONIA	Tree Fern	P	Spores	April	Greenhouse
DIEFFENBACHIA	Dumb Cane	P	Cuttings (stem sections)	June	Greenhouse/indoors
			Air layering	April–Aug	Greenhouse/indoors

Genus	Common name	Type	Method	Time	Location
DIMORPHOTHECA	Star of the Veld	An	Seeds	Feb–March	Greenhouse
		P	Semi-ripe cuttings	Aug	Greenhouse/frame
DIONAEA	Venus Fly Trap	P	Division	April	Greenhouse/indoors
			Seeds	March–April	Greenhouse
DIPLADENIA		Cl	Softwood cuttings	April–June	Greenhouse
DIZYGOTHECA		S	Seeds	Jan–April	Greenhouse/indoors
			Air layering	April–Aug	Greenhouse/indoors
DOROTHEANTHUS	Livingstone Daisy	An	Seeds	Feb–March	Greenhouse/frame
DOXANTHA (*Bignonia*)		Cl	Semi-ripe cuttings	June–Aug	Greenhouse
			Seeds	Jan–April	Greenhouse
DRACAENA		S T	Cuttings (stem sections)	June–July	Greenhouse/indoors
			Leaf-bud cuttings	June–July	Greenhouse/indoors
			Seeds	Jan–April	Greenhouse
			Air layering	April–Aug	Greenhouse/indoors
DROSERA	Sundew	P	Division	March–April	Greenhouse
			Seeds	March–April	Greenhouse
DURANTA		S	Softwood cuttings	April–May	Greenhouse
ECCREMOCARPUS	Chilean Glory Vine	Cl	Seeds	Feb	Greenhouse
ECHEVERIA		Suc	Leaf cuttings	June–Aug	Greenhouse/indoors
			Rooted offsets	April–Aug	Greenhouse/indoors
			Division	April	Greenhouse/indoors
			Seeds	Jan–April	Greenhouse/indoors

282

Plant		Type	Method	Time	Place
ECHINOCACTUS		Ca	Seeds	Jan–April	Greenhouse/indoors
ECHINOCEREUS		Ca	Seeds	Jan–April	Greenhouse/indoors
			Cuttings	June–July	Greenhouse/indoors
ECHINOPSIS		Ca	Seeds	Jan–April	Greenhouse/indoors
			Offsets	April	Greenhouse/indoors
EDGEWORTHIA		S	Semi-ripe cuttings	June–Aug	Greenhouse/frame
EICHHORNIA		Aq	Division	April–June	Greenhouse
EPACRIS		S	Semi-ripe cuttings	June–Aug	Greenhouse
EPIDENDRUM		Orch	Cuttings	After flowering	Greenhouse
			Division	When potting	Greenhouse
EPIPHYLLUM	Orchid Cactus	Ca	Cuttings (stem sections)	June–Aug	Greenhouse/indoors
			Seeds	Jan–April	Greenhouse/indoors
EPISCIA		P	Division	April–May	Greenhouse/indoors
			Cuttings	June–Aug	Greenhouse/indoors
ERICA	Cape Heath	S	Softwood cuttings	April–June	Greenhouse
ERYTHRINA	Coral Tree	T S	Seeds	Jan–April	Greenhouse
			Softwood cuttings	April–June	Greenhouse
EUCALYPTUS	Gum Tree	T	Seeds	March–April	Greenhouse

EUPHORBIA				
fulgens	S	Softwood cuttings	May–June	Greenhouse
pulcherrima Poinsettia	S	Softwood cuttings	May–June	Greenhouse
splendens Crown of Thorns	S	Softwood cuttings	May–June	Greenhouse
succulent species	Suc	Seeds	Jan–April	Greenhouse/indoors
		Cuttings	Spring–summer	Greenhouse/indoors
EXACUM	Bi (An)	Seeds	Feb–April	Greenhouse
			Sept–Oct	Greenhouse/frame
FAUCARIA Tigers Jaw	Suc	Seeds	Jan–April	Greenhouse/indoors
		Cuttings	June–Aug	Greenhouse/indoors
FEIJOA	S	Semi-ripe cuttings	July–Aug	Greenhouse
FELICIA	S	Softwood cuttings	May–June	Greenhouse
	An	Seeds	Feb–March	Greenhouse
FEROCACTUS	Ca	Seeds	Jan–April	Greenhouse/indoors
FICUS including Rubber Plants	T S	Soft or semi-ripe cuttings	April–July	Greenhouse
		Leaf-bud cuttings	April–July	Greenhouse
		Air layering	April–Aug	Greenhouse/indoors
		Seeds	Jan–April	Greenhouse
FITTONIA	P	Division	April	Greenhouse/indoors
		Cuttings	June–Aug	Greenhouse/indoors
FRANCOA	P	Seeds	Feb–March	Greenhouse
		Division	April	Greenhouse

Plant	Type	Method	Time	Place
FREESIA	Cm	Seeds	When ripe or March	Greenhouse
		Cormlets	Aug–Sept	Greenhouse/frame
FUCHSIA	S	Softwood or semi-ripe cuttings	April–Aug	Greenhouse/indoors/frame
GARDENIA	S	Softwood or semi-ripe cuttings	April–Aug	Greenhouse
GASTERIA	Suc	Division or offsets	April	Greenhouse/indoors
		Seeds	Jan–April	Greenhouse/indoors
		Leaf cuttings	April–Aug	Greenhouse/indoors
GAZANIA	P	Semi-ripe cuttings	Aug	Greenhouse/frame
		Seeds	Feb–March	Greenhouse
GERBERA Barberton Daisy	P	Division	April	Greenhouse
		Seeds	March–April	Greenhouse
		Cuttings	Spring or summer	Greenhouse
GESNERIA	P (Tu)	Division of tubers	When dormant	Greenhouse
		Seeds	Jan–April	Greenhouse
		Leaf cuttings	June–Aug	Greenhouse
GLADIOLUS	Cm	Cormlets	April	Outdoors
		Seeds	March–April	Greenhouse
		Division of corms	April	Outdoors

GLECHOMA	Ground Ivy	P	Division	April	Greenhouse/frame/indoors
			Cuttings	April–Aug	Greenhouse/frame/indoors
GLORIOSA	Climbing Lily	Tu	Seeds	Jan	Greenhouse
			Small tubers	Feb	Greenhouse
GOMPHOCARPUS		P S	Seeds	Jan–April	Greenhouse
GOMPHRENA	Globe Amaranth	An	Seeds	Feb–March	Greenhouse
GRAPTOPETALUM		Suc	Leaf cuttings	April–Aug	Greenhouse/indoors
GREVILLEA		T S	Seeds	Jan–April	Greenhouse
			Semi-ripe cuttings	June–July	Greenhouse
GUZMANIA		P	Division	May–July	Greenhouse/indoors
GYMNOCALYCIUM		Ca	Seeds	Jan–April	Greenhouse/indoors
			Offsets	June–Aug	Greenhouse/indoors
GYNURA		P	Cuttings	May–Aug	Greenhouse/indoors
HAEMANTHUS	Blood Lily	B	Bulblets	When new growth begins	Greenhouse
			Seeds	When ripe	Greenhouse
HAKEA		T S	Seeds	March–April	Greenhouse
			Semi-ripe cuttings	June–July	Greenhouse
HAWORTHIA		Suc	Division	July–Aug	Greenhouse/indoors
			Offsets	July–Aug	Greenhouse/indoors

Plant		Type	Method	Time	Place
HEDERA	Ivy	C1	Soft or semi-ripe cuttings	June–July	Greenhouse/indoors
			Layering (simple)	May–June	Greenhouse/indoors
			Leaf-bud cuttings	April–June	Greenhouse/indoors
HEDYCHIUM	Ginger Lily	P	Division	April	Greenhouse
HELIOTROPIUM	Heliotrope	S	Semi-ripe cuttings	Aug	Frame/greenhouse/indoors
			Softwood cuttings	April	Greenhouse
			Seeds	Feb–March	Greenhouse
HIBISCUS	Shrubby Mallow	S	Softwood cuttings	April–June	Greenhouse
HIPPEASTRUM		B	Bulblets	Autumn	Greenhouse/indoors
			Seeds	March	Greenhouse
HOWEA (Kentia)	Palm	T	Seeds	April	Greenhouse
HOYA	Wax Flower	C1	Semi-ripe cuttings	June–Aug	Greenhouse/indoors
HUMEA		Bi	Seeds	July	Frame
HYDRANGEA		S	Softwood cuttings	April	Greenhouse
HYMENOCALLIS		B	Bulblets	Sept–Oct	Greenhouse
HYPOCYRTA	Pouch Flower or Clog Plant	P S	Semi-ripe cuttings	June–Aug	Greenhouse/indoors
HYPOESTES	Polka Dot Plant	P	Seeds	Jan–April	Greenhouse

Genus	Common name	Type	Method	Time	Location
IMPATIENS	Balsam, Busy Lizzy	An P	Seeds Cuttings Seeds	March–April April–Aug March	Greenhouse/indoors Greenhouse/indoors Greenhouse/indoors
IPOMOEA	Morning Glory	Cl (An)	Seeds	March–April	Greenhouse
IRESINE		P	Soft cuttings	April–June	Greenhouse/indoors
JACARANDA		S T	Semi-ripe cuttings Seeds	June–July March–April	Greenhouse Greenhouse
JACOBINIA		S	Softwood cuttings	April–June	Greenhouse
JASMINUM	Jasmine	S Cl	Semi-ripe cuttings	Aug–Sept	Greenhouse
JOVELLANA		P	Semi-ripe cuttings	June–Aug	Greenhouse/frame
JUBAEA	Palm	T	Seeds	Jan–April	Greenhouse
KALANCHOE		Suc	Cuttings Leaf cuttings Seeds	June–Aug June–Aug Jan–April	Greenhouse/indoors Greenhouse/indoors Greenhouse/indoors
KOCHIA	Summer Cypress, Burning Bush	An	Seeds	March–April	Greenhouse
LACHENALIA	Cape Cowslip	B	Bulblets Seeds	Aug March–April	Greenhouse Greenhouse
LAELIA		Orch	Division	When potting	Greenhouse
x LAELIOCATTLEYA		Orch	Division	When potting	Greenhouse
LAGERSTROEMIA	Grape Myrtle	S T	Seeds Root cuttings Softwood cuttings	Autumn Dec–Jan May–June	Greenhouse Greenhouse Greenhouse

Plant	Type	Method	Time	Place
LAMPRANTHUS	Suc	Cuttings	Aug	Greenhouse/indoors
		Seeds	March–April	Greenhouse/indoors
LANTANA	S	Softwood cuttings	April–June	Greenhouse
		Seeds	Feb–March	Greenhouse
LAPAGERIA Chilean Bell Flower	Cl	Simple layering	April–June	Greenhouse
		Seeds	When ripe	Greenhouse
LAPEIROUSIA	Cm	Cormlets	Spring	Greenhouse
		Seeds	March–April	Greenhouse
LEONOTIS Lion's Ear	S	Semi-ripe cuttings	June–July	Greenhouse
LEPTOSPERMUM	S T	Semi-ripe cuttings	Aug	Greenhouse
		Seeds	March–April	Greenhouse
LEUCOCORYNE	B	Bulblets	Sept–Oct	Greenhouse
		Seeds	March–April	Greenhouse
LITHOPS Living Stones	Suc	Division	Summer	Greenhouse/indoors
		Seeds	Jan–April	Greenhouse/indoors
LOBELIA	P	Division	March–April	Greenhouse
	S	Soft or semi-ripe cuttings	April–Aug	Greenhouse
	P S	Seeds	March–April	Greenhouse
erinus	An	Seeds	Jan–March	Greenhouse
LOBIVIA	Ca	Seeds	Jan–April	Greenhouse/indoors
LOPHOPHORA Dumpling Cactus	Ca	Seeds	Jan–April	Greenhouse/indoors

LUCULIA		S	Semi-ripe cuttings	July–Aug	Greenhouse
LUFFA	Loofah	Cl (An)	Seeds	March–April	Greenhouse
LYCASTE		Orch	Division	Spring	Greenhouse
LYCOPODIUM	Club Moss	P	Spores Division	When ripe April	Greenhouse Greenhouse
MAMMILLARIA		Ca	Offsets (division) Seeds	Summer Jan–April	Greenhouse/indoors Greenhouse/indoors
MANDEVILLA	Chilean Jasmine	Cl	Seeds Semi-ripe cuttings	March–April June–July	Greenhouse Greenhouse
MANIHOT esculenta	Tapioca, Cassava, Manioc	S	Cuttings (tips or stem sections)	June–Aug	Greenhouse
MARANTA		P	Division Cuttings	April April	Greenhouse/indoors Greenhouse/indoors
MASDEVALLIA		Orch	Division	Spring or autumn	Greenhouse
MAXILLARIA		Orch	Division	Spring	Greenhouse
MEDINILLA		S	Soft or semi-ripe cuttings	April–Aug	Greenhouse
MELALEUCA	Bottle Brush	S	Softwood cuttings	May–June	Greenhouse
MELIA	Bead Tree	T	Seeds Semi-ripe cuttings	March–April June–July	Greenhouse Greenhouse

Plant	Type	Method	Time	Place
MELIANTHUS Honey Bush	S	Division (possibly)	April	Greenhouse
		Semi-ripe cuttings	June–Aug	Greenhouse
		Seeds	April	Greenhouse
MESEMBRYANTHEMUM	Suc	Seeds	March–April	Greenhouse/indoors
		Cuttings	June–Aug	Greenhouse/indoors
		Leaf cuttings	June–Aug	Greenhouse
METROSIDEROS	T S	Semi-ripe cuttings	June–July	Greenhouse
		Seeds	March–April	Greenhouse
MICHELIA	T	Simple layering	April–Aug	Greenhouse
		Semi-ripe cuttings	Aug–Sept	Greenhouse
MILTONIA	Orch	Division	Autumn or spring	Greenhouse
MIMOSA pudica Sensitive Plant	An	Seeds	March–April	Greenhouse/indoors
MIRABILIS jalapa Marvel of Peru	P	Seeds	Feb–March	Greenhouse
MITRARIA	S	Semi-ripe cuttings	June–Aug	Greenhouse
MONSTERA Swiss Cheese Plant	Cl	Cuttings	June–Aug	Greenhouse
		Leaf-bud cuttings	June–Aug	Greenhouse
		Simple layering	Spring or summer	Greenhouse/indoors

		Cm			
MORAEA		Cm	Cormlets Seeds	Sept Feb–March	Greenhouse Greenhouse
MUSA	Banana	P	Division Suckers or offsets	May–July May–July	Greenhouse/indoors Greenhouse/indoors
MYOSOTIDIUM		P	Seeds	March–April	Greenhouse
NELUMBO	Lotus	Aq	Division Seeds	April–June April	Greenhouse Greenhouse
NEMESIA		An	Seeds	Feb–April	Greenhouse
NEOREGELIA		P	Rooted offsets Seeds	April–July March–April	Greenhouse/indoors Greenhouse
NEPENTHES	Pitcher Plant	Cl	Cuttings (stem tips or laterals) Seeds	Spring Spring	Greenhouse (high temperature and humidity) Greenhouse (high temperature and humidity)
NEPHROLEPIS	Ladder Fern	F	Division	April	Greenhouse/indoors
NERINE	Guernsey Lily	B	Bulblets Seeds	Aug When ripe	Greenhouse Greenhouse
NERIUM	Oleander	S	Semi-ripe cuttings Seeds	Aug–Sept March–April	Greenhouse Greenhouse
NICOTIANA	Flowering Tobacco Plant	An	Seeds	Feb	Greenhouse
NIDULARIUM		P	Rooted offsets	April–July	Greenhouse
NOTOCACTUS		Ca	Seeds	Jan–April	Greenhouse/indoors

Plant		Type	Method	Time	Place
NYMPHAEA	Tropical Waterlilies	Aq	Division of tubers	Early summer	Greenhouse
OCHNA		S	Semi-ripe cuttings	July–Aug	Greenhouse
			Seeds	March–April	Greenhouse
ODONTOGLOSSUM		Orch	Division	Spring or autumn	Greenhouse
OLEA	Olive	T	Semi-ripe cuttings	July–Aug	Greenhouse
ONCIDIUM		Orch	Division	When potting	Greenhouse
OPLISMENUS		G (P)	Division	Summer	Greenhouse/indoors
			Cuttings	June–Aug	Greenhouse/indoors
OPUNTIA	Indian Fig, Prickly Pear	Ca	Cuttings of single pads	June–Aug	Greenhouse/indoors
			Seeds	Jan–April	Greenhouse/indoors
ORNITHOGALUM		B	Bulblets	Sept	Greenhouse
			Seeds	When ripe or Jan–Feb	Greenhouse/frame
OXALIS		P	Division	April	Greenhouse
			Seeds	March–April	Greenhouse
PANDANUS	Screw Pine	T S	Offsets or suckers	April–Aug	Greenhouse/indoors
			Seeds	Jan–April	Greenhouse
PANDOREA	Wonga-wonga Vine	Cl	Semi-ripe cuttings	July–Aug	Greenhouse
			Seeds	Jan–April	Greenhouse
PAPHIOPEDILUM	Lady's Slipper	Orch	Division	April–May	Greenhouse

		Ca	Seeds	Jan–April	Greenhouse/indoors
PARODIA					
PASSIFLORA	Passion Flower	Cl	Leaf-bud cuttings	May–June	Greenhouse
			Seeds	March–April	Greenhouse
			Serpentine layering	April–Aug	Greenhouse
PAVONIA		S	Soft or semi-ripe cuttings	April–Aug	Greenhouse
			Seeds	Jan–April	Greenhouse
PEDILANTHUS	Slipper Spurge	Suc	Cuttings	June–Aug	Greenhouse
PELARGONIUM		P S	Semi-ripe cuttings	Aug	Greenhouse/frame/indoors
x domesticum	Regal Pelargonium	P	Semi-ripe cuttings	Aug	Greenhouse/frame/indoors
x hortorum	Zonal Pelargonium	P	Semi-ripe cuttings	Aug	Greenhouse/frame/indoors
peltatum	Ivy-leaf Pelargonium	P	Semi-ripe cuttings	Aug	Greenhouse/frame/indoors
x hortorum (special seed-raised strains)	Zonal Pelargonium	P	Seeds	Dec–Jan	Greenhouse
PELLAEA	Cliffbrake	F	Division	April	Greenhouse/indoors
			Spores	When ripe	Greenhouse
PELLIONIA		P	Division	April	Greenhouse/indoors
			Cuttings	June–Aug	Greenhouse/indoors
PENTAS	Star Cluster	S P	Softwood cuttings	April–June	Greenhouse
			Seeds	March–April	Greenhouse

Plant		Type	Method	Time	Place
PEPEROMIA	Pepper Elder	P	Leaf cuttings	April–Aug	Greenhouse
			Division	April	Greenhouse/indoors
			Seeds	Jan–April	Greenhouse
PERESKIA		Ca	Cuttings	July–Aug	Greenhouse/indoors
			Seeds	Jan–April	Greenhouse/indoors
PERILLA		An	Seeds	Feb–March	Greenhouse
PERSEA	Avocado	T	Seeds	As soon as available	Greenhouse/indoors
PETUNIA		An	Seeds	Jan–March	Greenhouse
PHAIUS		Orch	Division	After flowering	Greenhouse
PHALAENOPSIS	Moth Orchid	Orch	Division (difficult)	Spring	Greenhouse
			Plantlets, if formed, on flowering stems	When rooted – remove and pot	Greenhouse
PHILODENDRON		S Cl	Cuttings – stem sections or tips	June–Aug	Greenhouse
			Leaf-bud cuttings	June–Aug	Greenhouse
			Air layering	April–Aug	Greenhouse/indoors
			Simple layering	April–Aug	Greenhouse/indoors
PHLOX drummondii		An	Seeds	Feb–April	Greenhouse
				Sept–Oct	Frame/greenhouse

PHOENIX	Palm, including Date Palm	T	Seeds Suckers	Jan–April June	Greenhouse Greenhouse/indoors
PHYSALIS		An	Seeds	Feb–March	Greenhouse
PILEA		P	Seeds Soft or semi-ripe cuttings Division	Jan–April April–Aug April	Greenhouse Greenhouse/indoors Greenhouse/indoors
PINGUICULA	Butterwort	P	Division Seeds	March–April March–April	Greenhouse Greenhouse
PISTIA	Water Lettuce	Aq	Division	April–June	Greenhouse
PITCAIRNIA		P	Offsets	April–July	Greenhouse
PITTOSPORUM		T S	Seeds Semi-ripe cuttings	March–April July–Aug	Greenhouse Greenhouse
PLATYCERIUM	Stag's Horn Fern	F	Division or plantlets Spores	April When ripe	Greenhouse/indoors Greenhouse
PLECTRANTHUS		P S	Seeds Cuttings Division	Jan–April June–Aug April	Greenhouse Greenhouse/indoors Greenhouse/indoors
PLEIONE		Orch	Division	After flowering	Greenhouse
PLUMBAGO		S	Soft or semi-ripe cuttings	May–July	Greenhouse

Plant		Type	Method	Time	Place
PLUMERIA	Frangipani	S T	Softwood cuttings	April	Greenhouse
POLIANTHES	Tuberose	B	Division	Autumn or spring	Greenhouse
POLYPODIUM		F	Division	March–April	Greenhouse/indoors
			Spores	When ripe	Greenhouse
POLYSCIAS		S T	Cuttings (stem sections or tips)	July–Aug	Greenhouse
PORTULACA		An	Seeds	Feb–March	Greenhouse
				April–May	Outdoors
PRIMULA					
x kewensis		An	Seeds	Feb–June	Greenhouse
malacoides		An	Seeds	May–Aug	Greenhouse/frame
obconica		An	Seeds	Feb–June	Greenhouse
sinensis		An	Seeds	May–July	Greenhouse/frame
PROSTANTHERA	Mint Bush	S	Semi-ripe cuttings	July–Aug	Greenhouse
PROTEA		S	Seeds	March–April	Greenhouse
			Semi-ripe cuttings	July–Aug	Greenhouse
PSEUDOPANAX		T S	Seeds	When ripe	Frame/greenhouse
PTERIS	Table Fern	F	Division	April	Greenhouse/indoors
			Spores	When ripe	Greenhouse
PUNICA	Pomegranate	S	Semi-ripe cuttings	June–July	Greenhouse
			Seeds	Jan–April	Greenhouse

Genus	Common name		Method	Timing	Location
QUAMOCLIT		An (Cl)	Seeds	Feb–March	Greenhouse
REBUTIA		Ca	Seeds Offsets	Jan–April Spring or summer	Greenhouse/indoors Greenhouse/indoors
REHMANNIA		P	Seeds Root cuttings	April–May Feb	Greenhouse Greenhouse
REINWARDTIA		S	Softwood cuttings	April–May	Greenhouse
RHAPHIOLEPIS		S	Semi-ripe cuttings	July–Aug	Greenhouse
RHIPSALIDOPSIS	Easter Cactus	Ca	Cuttings (one to three joints long)	June–Aug	Greenhouse/indoors
RHIPSALIS		Ca	Cuttings	June–Aug	Greenhouse/indoors
RHODODENDRON					
simsii	Indian Azalea	S	Semi-ripe cuttings	June–July	Greenhouse
other tender species and cultivars, including Javanese species		S	Seeds Simple layering Semi-ripe cuttings	March–April April–Aug June–July	Greenhouse Greenhouse Greenhouse
RHOEO	Boat Lily	P	Seeds Cuttings	Jan–April June–July	Greenhouse Greenhouse/indoors
RHOICISSUS	Grape Ivy	Cl	Semi-ripe cuttings Leaf-bud cuttings	June–Aug June–Aug	Greenhouse/indoors Greenhouse/indoors
RICINUS	Castor-oil Plant	An	Seeds	Jan–March	Greenhouse

Plant	Type	Method	Time	Place
ROCHEA	**Suc**	Cuttings	April	Greenhouse/indoors
		Seeds	March–April	Greenhouse
RUELLIA	**P S**	Division (if possible)	April	Greenhouse/indoors
		Cuttings	June–Aug	Greenhouse
		Seeds	Jan–April	Greenhouse
SAINTPAULIA African Violet	**P**	Division	April–May	Greenhouse/indoors
		Leaf cuttings	June–Aug	Greenhouse/indoors
		Seeds	Jan–April	Greenhouse
SALPIGLOSSIS Painted Tongue	**An**	Seeds	Jan–March	Greenhouse
			April–May	Outdoors
SALVIA splendens Scarlet Sage	**An**	Seeds	Jan–March	Greenhouse
other species Sage	**S**	Semi-ripe cuttings	June–July	Greenhouse/frame
		Seeds	March–April	Greenhouse
	P	Division	March–April	Greenhouse
		Seeds	March–April	Greenhouse
SANSEVIERIA	**P**	Leaf cuttings (sections)	June–Aug	Greenhouse/indoors
		Division (rooted suckers)	April	Greenhouse/indoors
trifasciata 'Laurentii'	**P**	Division (rooted suckers)	April	Greenhouse/indoors
SARRACENIA Pitcher Plant	**P**	Seeds	March–April	Greenhouse

		P (Tu)			
SAUROMATUM	Voodoo Lily	P (Tu)	Offsets	Spring	Greenhouse
SAXIFRAGA stolonifera	Mother of Thousands	P	Layering plantlets	Spring–summer	Greenhouse/indoors
SCHEFFLERA		S	Seeds Cuttings Air layering	Jan–April June–July April–Aug	Greenhouse Greenhouse Greenhouse/indoors
SCHIZANTHUS	Poor Man's Orchid	An	Seeds	Aug–Sept March–April April–May	Greenhouse/frame Greenhouse Outdoors
SCHLUMBERGERA	Christmas Cactus	Ca	Cuttings (two or three joints)	Summer	Greenhouse/indoors
SCINDAPSUS		Cl	Cuttings (stem sections or tips) Leaf-bud cuttings	June–Aug June–Aug	Greenhouse Greenhouse
SEDUM	Stonecrop	Suc	Seeds Cuttings Leaf cuttings Division	Jan–April June–July June–July Spring	Greenhouse/indoors Greenhouse/indoors Greenhouse/indoors Greenhouse/indoors
SELAGINELLA		P	Cuttings Division	Spring/summer April	Greenhouse Greenhouse/indoors
SELENICEREUS	Queen of the Night	Ca	Cuttings (stem sections)	June–Aug	Greenhouse/indoors

Plant		Type	Method	Time	Place
SENECIO		P S	Cuttings	Spring/summer	Greenhouse/indoors
		Suc			
cineraria	Cineraria	An	Seeds	Jan–April	Greenhouse/indoors
				Feb–March	Greenhouse
x hybridus		An	Seeds	April–Aug	Greenhouse
SETCREASEA		P	Division	April	Greenhouse/indoors
			Cuttings	April–Aug	Greenhouse/indoors
SINNINGIA	including Gloxinias	P Tu	Seeds	Jan–March	Greenhouse
			Leaf cuttings	June–Aug	Greenhouse
SMITHIANTHA		P	Seeds	Feb–March	Greenhouse
			Leaf cuttings	June–Aug	Greenhouse
SOLANUM					
capsicastrum	Winter Cherry	S	Seeds	Feb–March	Greenhouse
pseudocapsicum	Jerusalem Cherry	S	Seeds	Feb–March	Greenhouse
SOLEIROLIA (*Helxine*)	Mind-your-own-business	P	Division	Any time of year	Greenhouse/indoors
SOPHRONITIS		Orch	Division	Spring	Greenhouse
SPARMANNIA		S	Softwood cuttings	April–June	Greenhouse/indoors
			Seeds	March–April	Greenhouse
SPATHIPHYLLUM		P	Division	April	Greenhouse/indoors
SPREKELIA		B	Bulblets	Autumn	Greenhouse
			Seeds	March–April	Greenhouse

Genus	Common name	Type	Method	Time	Location
STAPELIA		Suc	Division	April	Greenhouse/indoors
			Seeds	Jan–April	Greenhouse/indoors
			Cuttings	Spring/summer	Greenhouse/indoors
STEPHANOTIS	Wax Flower	Cl	Seeds	Jan–April	Greenhouse
			Semi-ripe cuttings	July–Aug	Greenhouse/indoors
STRELITZIA	Bird of Paradise Flower	P	Division	April	Greenhouse
			Seeds	Jan–April	Greenhouse
STREPTOCARPUS	Cape Primrose	P	Seeds	Jan–March	Greenhouse
			Leaf cuttings (sections)	June–Aug	Greenhouse
STREPTOSOLEN		S	Softwood cuttings	May–June	Greenhouse
SYNGONIUM	Goose Foot	Cl	Cuttings (stem sections or tips)	June–Aug	Greenhouse
TAGETES	African and French Marigolds	An	Seeds	Feb–March	Greenhouse
TELOPEA		S	Seeds	Jan–April	Greenhouse
			Simple layering	March	Greenhouse
TETRANEMA	Mexican Foxglove	P	Seeds	Jan–April	Greenhouse
			Division	April	Greenhouse
THUNBERGIA		An P Cl S P	Seeds	Jan–April	Greenhouse
			Soft cuttings	June	Greenhouse

Plant		Type	Method	Time	Place
TIBOUCHINA	Glory Bush	S	Soft or semi-ripe cuttings	April–Aug	Greenhouse
TILLANDSIA		P	Division (suckers)	June	Greenhouse/indoors
TOLMIEA menziesii	Pick-a-back Plant	P	Leaf plantlets (peg down leaves)	Summer	Greenhouse/indoors
TORENIA		An	Seeds	Feb–April	Greenhouse
TRADESCANTIA	Wandering Jew	P	Cuttings Simple layering	April–Sept Spring–summer	Greenhouse/indoors Greenhouse/indoors
TRICHOCEREUS		Ca	Seeds	Jan–April	Greenhouse/indoors
TROPAEOLUM peregrinum	Canary Creeper	P (C1) An (C1)	Seeds Cuttings Seeds	Feb–April April–May Feb–April April–May	Greenhouse Greenhouse Greenhouse Outdoors
URSINIA		An P	Seeds Seeds	Feb–April April–May Feb–April	Greenhouse Outdoors Greenhouse
VALLOTA	Guernsey or Scarborough Lily	B	Bulblets Seeds	June–July March	Greenhouse Greenhouse

			Method	Time	Location
VANDA		Orch	Cuttings (stems with aerial roots)	Summer	Greenhouse
			Offshoots	Spring	Greenhouse
VELTHEIMIA		B	Bulblets	June–July	Greenhouse
			Entire leaves, as cuttings	When fully developed	Greenhouse
			Seeds	March	Greenhouse
x VENIDIO-ARCTOTIS		P	Semi-ripe cuttings	Aug	Greenhouse
			Softwood cuttings	April	Greenhouse
VENIDIUM	Monarch of the Veldt	An	Seeds	Feb–March	Greenhouse
VERBENA	Vervain	An	Seeds	Jan–March	Greenhouse
		P	Seeds	Jan–March	Greenhouse
			Semi-ripe cuttings	Aug	Greenhouse/frame
VRIESIA		P	Division (offsets)	June–Aug	Greenhouse/indoors
WATSONIA		Cm	Division of clumps	Spring or late summer	Greenhouse
			Seeds	Jan–Feb	Greenhouse
WOODWARDIA radicans	Chain Fern	F	Bulbils on fronds	When available	Greenhouse
ZANTEDESCHIA	Arum Lily	P	Seeds	March–April	Greenhouse
			Division (offsets)	Spring	Greenhouse

Plant		Type	Method	Time	Place
ZEA					
mays	Ornamental Maize	An	Seeds	March–April	Greenhouse
				May	Outdoors
ZEBRINA	Wandering Jew	P	Cuttings	April–Aug	Greenhouse/indoors
			Simple layering	Spring/ summer	Greenhouse/indoors
ZEPHYRANTHES	Zephyr Lily	B	Bulblets	April	Greenhouse
			Division of clumps	April	Greenhouse
			Seeds	Jan–Feb	Greenhouse
ZINNIA		An	Seeds	March–April	Greenhouse
				May	Outdoors

Select bibliography

BROWN, GEORGE E., *The Pruning of Trees, Shrubs and Conifers*. Faber, 1972.

GARNER, R.J., *The Grafter's Handbook*. Faber, 1947.

HADFIELD, MILES, *A History of British Gardening*. John Murray, 1979.

HAY, R., MCQUOWN, F.R. & BECKETT, G. & K., *The Dictionary of Indoor Plants in Colour*. Michael Joseph, 1975.

HAY, R. & SYNGE, P., *The Dictionary of Garden Plants in Colour*. Michael Joseph, 1969.

HUXLEY, ANTHONY, *An Illustrated History of Gardening*. Paddington Press, 1978.

LEMON, KENNETH, *The Golden Age of Plant Hunters*. Phoenix (Dent), 1968.

LEMON, KENNETH, *The Covered Garden*, Museum Press, 1972.

ROYAL HORTICULTURAL SOCIETY, *The Dictionary of Gardening*. Oxford University Press, 1951; Supplement, 1969.

THOMAS, GRAHAM STUART, *Perennial Garden Plants*. Dent, 1976.

THOMAS, GRAHAM STUART, *Plants for Ground Cover*. Dent, 1970; Revised 1977.

Index

General Index

Bell glass, 13, 14, 15
Budding, 16, 19, 119, 121, 122, 124; roses, 119, 121, 122, 123 (Fig. 31), 124; rose rootstocks for, 120, 121; trees, 124
Bulblets, 137 (Fig. 36), 138
Bulbils, 138 (Fig. 37), 139
Compost, potting, 59, 166, 167; seed, 54
Cormlets, 141 (Fig. 39), 142
Cuttings, 13, 14, 15, 18 basal, 63
  eye, 97; insertion of, 97, 98, 99 (Fig. 24); preparation of, 97, 98 (Fig. 23)
  hardwood, 77; collection of, 77; insertion of, 79 (Fig. 16), 80, 81; preparation of, 78 (Fig. 15)
  Irishman's, 102
  leaf, 22, 89; insertion of, 89, 90 (Fig. 19), 91, 92 (Fig. 20), 93 (Fig. 21); preparation of, 89, 90 (Fig. 19), 91, 92 (Fig. 20), 93 (Fig. 21); rooting, 91
  leaf-bud, 94, 95, 96 (Fig. 22)
  mallet, 71, 72 (Fig. 12)
  root, 22, 83; collection of, 83, 84; insertion of, 86, 87 (Fig. 18); preparation of, 84, 85 (Fig. 17); rooting, 85, 86
  rooted, aftercare of, 69, 76, 77, 81, 82, 86, 87, 88, 91, 94, 99, 101, 102
  rooting in water, 101, 102
  semi-ripe, 69; collection of, 70; insertion of, 73; preparation of, 70, 71

(Fig. 11), 73 (Fig. 13); rooting, 74, 75, 76
  softwood, 63; collection of, 63, 64; insertion of, 66, 67 (Fig. 10); preparation of, 64, 65 (Fig. 8), 66 (Fig. 9); rooting, 66, 67, 68, 69 wounding, 70
Dicing rhizomes of Bergenia, 129, 130
Division, 19; of alpines, 130, 131 (Fig. 34), 132; of aquatics, 132, 133, 134 (Fig. 35); of bulbs, 136, 137 (Fig. 36); of corms, 136, 141 (Fig. 39); of ferns, 130; of grasses, 130; of greenhouse plants, 135; of hardy perennials, 125, 126, 127 (Fig. 32), 128 (Fig. 33); of orchids, 135; of shrubs, 136; of tubers, 136, 142 (Fig. 40), 143
Grafting, 13, 15, 16, 19, 22, 103; purpose of, 103; rootstocks for, 105, 106, 110, 112, 114, 115, 117; saddle, 109, 110, 111 (Fig. 27), 112; spliced side, 112, 113 (Fig. 28); spliced side veneer, 115, 117, 118 (Fig. 30); veneer, 112, 114, 115, 116 (Fig. 29); whip-and-tongue, 104, 105, 106, 107 (Fig. 26), 108
Growing rooms, 17
Hardening plants, 23, 169
Knives, 104
Layering, 13, 17, 19, 22 air, 17, 22, 150, 151 (Fig. 42), 153 (Fig. 43); greenhouse plants, 152, 154; trees and shrubs, 150, 151, 152

carnation, 160, 161 (Fig. 47), 162
by dropping, 157
greenhouse plants, 162, 163 (Fig. 48), 164
serpentine, 154, 155 (Fig. 44)
simple, 144, 145, 146, 147, 148 (Fig. 41), 149
strawberry runners, 158, 159 (Fig. 46), 160
tip, 156 (Fig. 45), 157
Lifting plants, 172, 174, 175
Nursery,
  management, 165; planting, 170, 171; site for, 169, 170
Pipings, 100 (Fig. 25), 101
Polythene tunnel, 74, 75 (Fig. 14), 76
Pots, 166
Potting, 22, 165, 166, 167, 168 (Fig. 49)
Potting on, 167
Pricking out seedlings, 22, 57, 58 (Fig. 7), 59, 61
Propagating case, 15, 67
Propagation, mist, 15, 67, 68
Scales, bulb, 22; lily, 139, 140 (Fig. 38), 141
Seeds, 13, 18, 21; alpines from, 37; annuals, half-hardy, from, 59; annuals, hardy, from, 47; biennials, hardy, from, 50; bulbs from, 42; collecting, 25, 26, 27, 28, 38, 42, 44; conifers from, 25; corms from, 42; drying and cleaning, 28, 29, 38, 43, 44; grasses from, 44; greenhouse plants from, 53; house plants from, 53; perennials, half-hardy, from, 59; perennials,

Seeds (*cont.*)
  hardy, from, 44; shrubs
  from, 25; sowing, 33, 34,
  35 (Fig. 3), 36, 39, 40, 41
  (Fig. 4), 43, 45, 48, 49, 51,
  53, 54 (Fig. 5), 55, 56
  (Fig. 6), 57, 59, 60;
  storing, 29, 30 (Fig. 1),
31, 38, 39, 43, 44;
  stratification of, 31, 32
  (Fig. 2), 33; trees from, 25
Soil-warming cables, 68
Spores, 21, 60; greenhouse
  ferns from, 60; hardy
  ferns from, 46, 47
Suckers, 136

Tissue culture, 17
Training plants, 22, 171,
  172; trees, 108, 109
Trimming plants, 171,
  172, 173 (Fig. 50)
Young plants, care of, 36,
  37, 40, 41, 43, 45, 46, 50,
  52, 60, 61

Index of Plant Names

Abelia, 180
Abeliophyllum, 180
Abies, 180
Abutilon, 180, 268
Acacia, 268
Acacia, False, 105, 251
Acaena, 180
Acalyphan, 268
Acantholimon, 180
Acanthopanax, 180
Acanthus, 180
Acer, 27, 28, 30, 64, 180
  campestre, 15
  japonicum, 117, 170, 180
  palmatum, 117, 170, 180
  platanoides, 105, 106
  pseudoplatanus, 105, 106
Achillea, 181
Achimenes, 268
Acidanthera, 268
Aconitum, 181
Acorus, 181
Actaea, 181
Actinidia, 181
  kolomikta, 81
Adenophora, 181
Adiantum, 60, 181, 268
Adonis, 39, 181
Aechmea, 268
Aegopodium, 181
Aeonium, 268
Aerides, 268
Aeschynanthus, 268
Aesculus, 33, 181
Aethionema, 182
African Violet, 298
Agapanthus, 182
Agapetes, 268
Agastache, 182
Agave, 269

Ageratum, 269
Aglaonema, 269
Agrimonia, 182
Agrimony, 182
Agrostemma, 182
Aichryson, 269
Ailanthus, 182
Ajuga, 182
Akebia, 182
Alchemilla, 182
Alder, 183
Alisma, 182
Alkanet, 184
Allamanda, 269
Allium, 138, 182
Almond, 14, 105, 248
Alnus, 183
Aloe, 91, 269
Alonsoa, 269
Alopecurus, 183
Alstroemeria, 126, 183
Alternanthera, 269
Althaea, 183
Alum, 218
Alyssum, 183
Alyssum, Sweet, 228
Amaranthus, 269
Amaryllis, 183
Amelanchier, 183
Ammobium, 269
Amorpha, 183
Ampelopsis, 183
Amsonia, 183
Anacyclus, 184
Anagallis, 184
Ananas, 269
Anaphalis, 184
Anchusa, 128, 184
Andromeda, 184
Andromischus, 268

Androsace, 38, 39, 184
  sarmentosa, 131
Anemone, 39, 184
  × *hybrida*, 126, 184
  Japanese, 126, 184
Anemonopsis, 184
Angelica, 184
Angel's Trumpets, 280
Angraecum, 269
Anigozanthus, 269
Aniseed, 245
Antennaria, 184
Anthemis, 185
Anthericum, 185
Antholyza, 270
Anthriscus, 185
Anthurium, 270
Anthyllis, 185
Antirrhinum, 270
Aphelandra, 270
Aponogeton, 185
Aporocactus, 270
Apple, 14, 16, 103, 105,
  231
Apple, Crab, 28, 31, 105,
  231
Apple, Malling Merton
  and Malling Root-
  stocks, 105, 106
Apple of Peru, 236
Apricot, 105, 248
Aquilegia, 20, 44, 185
Arabis, 185
Aralia, 85, 185
Araucaria araucana, 185
  excelsa, 270
Arbor-vitae, 261
  Hiba, 261
Arbutus, 33, 34, 185
Arctostaphylos, 185

Index

Arctotis, 270
Ardisia, 270
Arenaria, 186
Argemone, 186
Arisarum, 186
Aristolochia, 186
Armeria, 186
Arnebia, 186
Aronia, 186
Arrowhead, 253
Artemisia, 186
Arum, 186
Aruncus, 186
Arundinaria, 186
Arundo, 186
Asarum, 186
Asclepias, 187, 270
Ash, 15, 26, 27, 28, 30, 33,
  212
  Mountain, 31, 105, 257
  Prickly, 267
Asparagus, 270
Asperula, 187
Asphodel, 187
  Giant, 209
Asphodeline, 187
Aspidistra, 135, 271
Asplenium, 187, 271
Aster, 102, 187
Astilbe, 187
Astragalus, 187
Astrantia, 187
Astrophytum, 271
Asystasia, 271
Athrotaxis, 187
Athyrium, 187
Atriplex, 187
Aubrieta, 37, 187
Aucuba, 188
Auricula, 20, 247
Avena, 188
Avens, 214
Avocado, 294
Azalea, Deciduous, 64,
  251
  Evergreen, 251
  Indian, 297
Azara, 188
Azolla, 188

Babiana, 271
Baby Blue Eyes, 236

Ballota, 188
Balm, Bastard, 232
  Bee, 234
  Lemon, 232
Balsam, 286
Bamboo, 186, 244
Banana, 291
Banksia, 271
Baptisia, 188
Barbarea, 188
Barberry, 188
Barosma, 271
Barrenwort, 208
Bartonia, 233
Basil, 13, 14
Bay, 14, 224
Bead Tree, 289
Beard Tongue, 241
Bear's Breeches, 180
Beauty Bush, 223
Bedstraw, 213
Beech, 15, 26, 30, 105, 117,
  144, 211
  Southern, 237
Beet, 13
Begonia, 55, 84, 271
  masoniana, 89, 271
  rex, 89, 271
Bellflower, 192, 273
  Ring, 259
Bellis, 51, 188
Bells of Ireland, 234
Bellwort, 263
Beloperóne, 271
Berberidopsis, 188
Berberis, 31, 33, 70, 71, 72,
  74, 76, 172, 188
  'Chenaultii', 72
  gagnepainii, 72
  × ottawensis, 72
  × stenophylla, 72
  'Irwinii', 71
  thunbergii, 71
  wilsoniae, 71
Bergamot, 234
Bergenia, 44, 129, 130, 189
Beschorneria, 189
Betula, 117, 189
  pendula, 26, 117
Bignonia, 281
Bilberry, 264
Billardiera, 272

Billbergia, 272
Birch, 117, 189
  Silver, 26, 117
Bird of Paradise
  Flower, 301
Blackberry, 95, 156, 157,
  252
Bladder Nut, 257
Bladder Senna, 199
Bladderwort, 263
Blanket Flower, 213
Blechnum, 189, 272
Bletia, 272
Bletilla, 189, 272
Bloodroot, 253
Bluebell, 208
Blueberry, 264
Blue Oxalis, 241
Bluets, 219
Bog Bean, 233
Bolax, 189
Bomarea, 272
Borage, 189
Borago, 189
Border Phlox, 243
Boronia, 272
Borzicactus, 272
Bottlebrush, 272, 289
Bougainvillea, 272
Bouvardia, 272
Box, 26, 191
Box Thorn, 230
Boykinia, 189
Brachycome, 189
Bramble, 17, 144
  Ornamental, 252
Brassia, 272
× Brassocattleya, 272
Briar, Wild, 120
Briza, 190
Brodiaea, 190
Bromus, 190
Broom, 26, 27, 203, 214
  Hedgehog, 209
  Southern, 237
  Spanish, 257
Browallia, 272
Bruckenthalia, 190
Brunfelsia, 273
Brunnera, 190
Bryophyllum, 21, 91, 273
Buckthorn, 26, 250

Buddleia, 26, 81, 82, 172, 190
Bugbane, 197
Bugle, 182
Bulbinella, 190
Bulbocodium, 190
Buphthalmum, 190
Bupleurum, 191
Burnet, 252
Burning Bush, 205, 287
Bush Clover, 225
Busy Lizzy, 286
Butcher's Broom, 253
Butomus, 191
Butter Bur, 242
Buttercup, 250
  Water, 250
Butterwort, 245, 295
Buxus, 191
Buxus sempervirens, 26

Cabbage, 13
Cactus, Christmas, 299
  Dumpling, 288
  Easter, 297
  Old Man, 274
  Orchid, 282
  Peanut, 275
  Rat-tail, 270
  Star, 271
Caesalpinia, 273
Caladium, 273
Calamint, 191
Calamintha, 191
Calamondin, 276
Calandrina, 273
Calanthe, 273
Calathea, 273
Calceolaria, 191, 273
Calendula, 191
California Fuchsia, 267
Calla, 191
Callicarpa, 191
Callirrhoe, 191
Callisia, 273
Callistemon, 273
Callistephus, 273
Callitriche, 191
Calluna, 20, 34, 36, 70, 157, 166, 191
  vulgaris, 26
Calocedrus, 192

Calochortus, 192
Caltha, 192
Calycanthus, 192
Camassia, 192
Camellia, 19, 34, 36, 94, 95, 166, 170, 192
Campanula, 51, 192
  isophylla, 273
  medium, 192
Campion, 230, 232, 256
Campsis, 193
Canary Creeper, 302
Candytuft, 221
  Perennial, 221
Canna, 176, 274
Canterbury Bell, 51, 192
Cape Figwort, 243
Cape Primrose, 301
Capsicum, 274
Caragana, 193
Cardamine, 193
Cardiocrinum, 193
Carduncellus, 193
Carex, 193
Carlina, 193
Carnation, 20, 128, 150, 160, 204
  Annual, 205
  Perpetual, 280
Carnegiea, 274
Carpenteria, 193
Carpinus, 193
Carya, 194
Caryopteris, 64, 194
Cassava, 289
Cassia, 274
Cassinia, 194
Cassiope, 194
Castanea, 33, 194
Cast-iron Plant, 271
Castor-oil Plant, 297
Catalpa, 194
Catananche, 194
Catchfly, 256
Catmint, 236
Cattleya, 274
Ceanothus, 86, 172, 194
Cedar, 114, 194
  Incense, 192
Cedrus, 114, 194
  atlantica glauca, 114
  deodara, 114

Celastrus, 195
  orbiculatus, 81
Celery, 13, 14
Celmisia, 195
Celosia, 274
Celsia, 274
Centaurea, 195, 274
Centranthus, 195
Century Plant, 269
Cephalaria, 195
Cephalocereus, 274
Cephalotaxus, 195
Cerastium, 195
Ceratostigma, 195
Cercidiphyllum, 195
Cercis, 196
Cereus, 274
Ceropegia, 274
Cestrum, 274
Chaenomeles, 196
Chamaecereus, 275
Chamaecyparis, 196
  lawsoniana, 114
  obtusa, 114
Chamaedaphne, 196
Chamaedorea, 275
Chamaemelum, 196
Chamaepericlymenum canadense, 200
Chamaerops, 275
Chamomile, 185, 196
Cheiranthus, 18, 51, 196
Chelone, 196
Cherry, 31, 103, 105, 248
  Dwarf, 248
  Wild, 105
Chervil, 185
Chestnut, Horse, 27, 33, 181
  Sweet, 27, 33, 194
Chiastophyllum, 196
Chilean Bell Flower, 288
Chilean Fire Bush, 208
Chimonanthus, 197
China Aster, 273
Chinese Bellflower, 245
Chinese Lantern, 244
Chionanthus, 197
Chionodoxa, 42, 197
Chlidanthus, 275
Chlorophytum, 135, 162, 163, 275

Index

Choisya, 197
Chokeberry, 186
Chorizema, 275
Christmas Rose, 218
Chrysanthemum, 63, 102, 197, 275
  Annual, 197
  Cascade, 275
  Charm, 275
  Early-flowering Border, 197
  frutescens, 275
  Korean, 275
  Late-flowering, 275
  parthenium, 275
Chrysogonum, 197
Cimicifuga, 197
Cineraria, 275
Cineraria, 300
Cinquefoil, 247
Circium, 197
Cissus, 275
Cistus, 198
Citron, 276
Citrus, 14, 16, 276
Cladanthus, 198
Cladrastis, 198
Clarkia, 198
Clary, 253
Cleistocactus, 276
Clematis, 28, 94, 154, 198
  vitalba, 28
Cleome, 276
Clerodendron, 136, 198, 276
Clethra, 198, 276
Clianthus, 276
Cliff-brake, 293
Clivia, 276
Clog Plant, 286
Clover, 262
Club Moss, 289
Clubrush, 255
Cobaea, 276
Cockscomb, 274
Cocksfoot, 203
Cocos, 276
Codiaeum, 101, 152, 277
Codonopsis, 198
Coelogyne, 277
Coffea, 277
Coffee, 277

Coix, 199
Colchicum, 199
Coleus, 101, 277
Colletia, 199
Collinsia, 199
Colquhounia, 277
Columbine, 185
Columnea, 277
Colutea, 27, 34, 199
Comfrey, 259
Commelina, 277
Coneflower, 252
Conophytum, 277
Convallaria, 199
Convolvulus, 199
Copiapoa, 277
Coprosma, 199, 277
Coral Tree, 282
Cordyline, 278
Coreopsis, 199
Coriander, 13
Coriaria, 199
Cork Tree, 242
Corn Cockle, 182
Cornel, 200
Cornflower, 195
Cornus, 81, 136, 172, 200
  canadensis, 200
  sanguinea, 26
Corokia, 200
Coronilla, 200, 278
Cortaderia, 130, 200
Corydalis, 39, 200
Corylopsis, 200
Corylus, 200
Coryphantha, 278
Cosmea, 278
Cosmos, 278
  atrosanguineus, 278
Cotinus, 68, 200
Cotoneaster, 28, 31, 33, 144, 172, 201
Cotula, 201, 278
Cotyledon, 278
Cowslip, 247
  Cape, 287
Crambe, 201
Crane's Bill, 214
Crassula, 91, 201, 278
+ *Crataegomespilus*, 201
Crataegus, 33, 105, 201
  monogyna, 105, 106

Crepis, 201
Cress, 13
Crinodendron, 201
× *Crinodonna*, 278
Crinum, 201, 278
Crocosmia, 202
Crocus, 42, 141, 142, 202
Crocus, Chilean, 260
Crossandra, 279
Croton, 152, 277
Crow Berry, 208
Crowfoot, 250
  Water, 250
Crown of Thorns, 283
Cryptanthus, 279
Cryptogramma, 202
Cryptomeria, 202
Ctenanthe, 279
Cucumber, 13
Cucurbita, 279
Cunninghamia, 202
Cup Flower, 236
Cuphea, 279
× *Cupressocyparis*, 177, 202
  leylandii, 171
Cupressus, 114, 171, 202
  glabra 'Pyramidalis', 114
  macrocarpa, 114
Currant, Black, 80, 251
  Flowering, 251
  Red, 78, 80, 251
  White, 251
Curtonus, 202
Cyananthus, 202
Cyanotis, 279
Cycas, 279
Cyclamen, 42, 202, 279
Cydonia, 202
  oblonga, 105
Cymbalaria, 203
Cymbidium, 135, 279
Cynoglossum, 203
Cyperus, 279
Cyphomandra, 280
Cypress, 13, 14, 202
  False, 196
  Leyland, 171
  Swamp, 260
Cypripedium, 280
  calceolus, 203
Cyrtomium, 280
Cystopteris, 203

Cytisus, 27, 34, 37, 171,
   203, 280
  purpureus, 178
  scoparius, 26

Daboecia, 157, 166, 203
Dactylis, 203
Daffodil, 136, 137, 236
Dahlia, 63, 142, 143, 176,
   280
Daisy, 188
  African, 229, 270
  Barberton, 284
  Double-flowered, 51, 188
  Globe, 215
  Livingstone, 281
  Michaelmas, 102, 187
  Ox-eye, 197
  Shasta, 197
  Swan River, 189
Daisy Bush, 238
Damson, 248
Danaë, 203
Daphne, 203
Darlingtonia, 204
Datura, 280
Davidia, 204
Decaisnea, 204
Delphinium, 20, 44, 45, 204
Dendrobium, 135, 280
Dendromecon, 204
Deodar, 114
Deschampsia, 204
Desfontainea, 204
Deutzia, 81, 172, 204
Dianthus, 20, 51, 100, 128,
   150, 160, 204, 280
  barbatus, 205
Dicentra, 205
Dicksonia, 280
Dictamnus, 205
Didiscus, 205
Dieffenbachia, 152, 280
Dierama, 205
Diervilla, 172, 205
Digitalis, 51, 205
Dill, 13
Dimorphotheca, 281
Dionaea, 281
Dionysia, 205
Dipelta, 205
Dipladenia, 281

Dipsacus, 205
Disanthus, 205
Disporum, 205
Distylium, 206
Dizygotheca, 281
Dodecatheon, 206
Dogwood, 26, 81, 172, 200
  Creeping, 200
Doronicum, 126, 206
Dorotheanthus, 281
Dorycnium, 206
Douglasia, 206
Doxantha, 281
Draba, 206
Dracaena, 95, 152, 281
Dracocephalum, 206
Dracunculus, 206
Dragon's Head, 206
Dragon Plant, 206
Drimys, 206
Drosera, 207, 281
Dryas, 207
Dryopteris, 207
Duckweed, 225
Dumb Cane, 152, 280
Duranta, 281

Earth Star, 279
Eccremocarpus, 281
Echeveria, 91, 281
Echinacea, 207
Echinocactus, 282
Echinocereus, 282
Echinops, 126, 207
Echinopsis, 282
Echium, 207
Edelweiss, 225
Edgworthia, 282
Edraianthus, 207
Eichhornia, 282
Elaeagnus, 71, 76, 207
Elder, 253
Elm, 263
Elodea, 207
Elsholtzia, 207
Elymus, 207
Embothrium, 208
Empetrum, 208
Endymion, 208
Enkianthus, 208
Eomecon, 208
Epacris, 282

Epidendrum, 282
Epigaea, 208
Epilobium, 208
Epimedium, 126, 208
Epiphyllum, 282
Episcia, 282
Eranthis, 208
Eremurus, 209
Erica, 20, 34, 36, 70, 157,
   166, 209, 282
Erigeron, 209
Erinacea, 209
Erinus, 209
Eriobotrya, 209
Eritrichium, 209
Erodium, 209
Eryngium, 85, 126, 209
Erysimum, 209
Erythrina, 282
Erythronium, 209
Escallonia, 210
Eschscholzia, 210, 282
Eucalyptus, 210, 282
Eucomis, 210
Eucryphia, 210
Euonymus, 33, 210
  radicans, 210
Eupatorium, 210
Euphorbia, 210, 283
  fulgens, 283
  marginata, 211
  pulcherrima, 283
  splendens, 283
Euryops, 211
Evening Primrose, 237
Exacum, 283
Exochorda, 211

Fabiana, 211
Fagus, 30, 144, 211
  sylvatica, 15, 26, 105,
   106, 117
False Dragonhead, 244
× *Fatshedera*, 211
Fatsia, 211
Faucaria, 283
Feijoa, 283
Felicia, 283
Fennel, 211
Fern, 21
  Bladder, 203
  Buckler, 207

Index

Fern *(cont.)*
  Chain, 266
  Hart's Tongue, 244
  Holly, 246
  Ladder, 291
  Maidenhair, 181, 268
  Mountain Parsley, 202
  Ostrich, 232
  Royal, 239
  Sensitive, 238
  Shield, 246
  Stag's Horn, 295
  Table, 296
  Tree, 280
Ferocactus, 283
Fescue, 211
Festuca, 211
Ficus, 152, 283
  carica, 81, 211
  elastica, 95, 152
Fig, 14, 81, 211
  Indian, 292
Filbert, 200
Filipendula, 211
Fir, Douglas, 248
Fire Thorn, 249
Fittonia, 283
Fitzroya, 211
Flax, 228
Flea Bane, 209
Floss Flower, 269
Foam Flower, 261
Foeniculum, 211
Fontanesia, 212
Forget-me-not, 51, 235
Forget-me-not,
  Water, 235
Forsythia, 81, 172, 212
  suspensa, 144
Fothergilla, 212
Foxglove, 51, 205
Foxglove, Fairy, 209
Foxglove, Mountain, 239
Foxtail, 183
Foxtail Barley, 219
Fragaria, 20, 150, 158, 212
Frankenia, 212
Francoa, 283
Frangipani, 296
Fraxinus, 28, 30, 33, 212
  excelsior, 15, 26
Freesia, 284

Fremontia, 212
Fritillaria, 212
Fritillary, 212
Frogbit, 220
Fuchsia, 63, 101, 212, 284

Galium, 213
Galtonia, 213
Gardenia, 284
Garlic, 138
Garrya, 213
  elliptica, 170
Gasteria, 284
Gaultheria, 157, 166, 213
  × *Gaulnettya*, 214
Gaura, 214
Gazania, 176, 284
Genista, 27, 34, 37, 171, 214
Gentian, 214
Gentiana, 214
Geranium, 214
Gerbera, 284
Germander, 260
Gesneria, 284
Geum, 214
Gilia, 214
Gillenia, 214
Ginkgo, 214
Gladiolus, 141, 142, 176, 284
Glaucium, 215
Glechoma, 215, 285
Gleditsia, 215
Globe Amaranth, 285
Globe Flower, 263
Globularia, 215
Gloriosa, 285
Glory Bush, 302
Gloxinia, 89, 300
Glyceria, 215
Glycyrrhiza, 215
Goat's Beard, 186
Goat's Beard, False, 187
Godetia, 215
Golden Bell, 212
Golden Club, 239
Golden Rain Tree, 223
Golden Rod, 256
Gomphocarpus, 285
Gomphrena, 285
Gooseberry, 78, 80, 251
Goose Foot, 301

Gorse, 26, 214, 263
Gourd, Ornamental, 279
Grapefruit, 276
Grape Ivy, 297
Graptopetalum, 285
Grass, Brome, 190
  Hair, 204
  Hare's Tail, 224
  Lyme, 207
  Ornamental, 44
  Pampas, 130, 200
  Quaking, 190
  Ribbon, 242
  Squirrel Tail, 219
  Whitlow, 206
  Zebra, 130
Greater Spearwort, 250
Grevillea, 215, 285
Griselinia, 215
Gromwell, 228
Ground Ivy, 215, 285
Guelder Rose, 26
Gum Tree, 210, 282
Gunnera, 215
Guzmania, 285
Gymnocalycium, 285
Gymnocladus, 216
Gynura, 285
Gypsophila, 216

Haberlea, 89, 216
Hacquetia, 216
Haemanthus, 285
Hakea, 285
Halesia, 216
× *Halimiocistus*, 216
Halimium, 216
Hamamelis, 19, 34, 36, 216
Handkerchief Tree, 204
Haplopappus, 217
Hare's Ear, 191
Hawkweed, 201, 218
Hawthorn, 105
  Water, 185
Haworthia, 285
Hazel, 200
Hearts Entangled, 274
Heath, 20, 70, 74, 76, 157, 209
  Cape, 282
  St Dabeoc's, 203
  Sea, 212

Heather, 20, 26, 70, 74, 76, 157, 191, 209
Heavenly Bamboo, 236
Hebe, 172, 217
Hedera, 94, 164, 217, 286
  helix, 15
Hedychium, 286
Helenium, 217
Helianthemum, 217
Helianthus, 217
Helichrysum, 217
Helictotrichon, 217
Heliopsis, 217
Heliotrope, 286
Heliotropium, 286
Helipterum, 218
Hellebore, 44, 218
  False, 264
Helleborus, 126, 218
Helxine, 300
Hemerocallis, 218
Hemlock, 263
Hemp Agrimony, 210
Hepatica, 218
Hesperis, 218
Heuchera, 218
× *Heucherella*, 218
Hibiscus, 218, 286
Hickory, 194
Hieraceum, 218
Hippeastrum, 286
Hippophae, 219
Hoheria, 219
Holcus, 219
Holly, 28, 31, 71, 76, 221
Hollyhock, 183
Holodiscus, 219
Honesty, 51, 229
Honey Bush, 290
Honeysuckle, 154, 229
  Himalayan, 226
Hop, 219
Hop Tree, 248
Hordeum, 219
Horminum, 219
Hornbeam, 193
Hornbeam, Hop, 239
Horned Rampion, 244
Hosta, 219
Hottonia, 219
Hound's Tongue, 203
Houseleek, 131, 255

Houstonia, 219
Houttuynia, 219
Howea, 286
Hoya, 286
Humea, 286
Humulus, 219
Hutchinsia, 219
Hyacinth, 136, 220
Hyacinth, Grape, 235
Hyacinthus, 136, 220
Hydrangea, 166, 220, 286
  Climbing, 220
  petiolaris, 220
Hydrocharis, 220
Hymenanthera, 220
Hymenocallis, 286
Hypericum, 27, 34, 220
Hypocyrta, 286
Hypoestes, 286
Hypoxis, 220
Hyssop, 221
Hyssopus, 221

Iberis, 221
Idesia, 221
Ilex, 28, 31, 33, 71, 76, 221
Immortelle, 266
  Sand, 269
Impatiens, 101, 286
Incarvillea, 221
Indian Bean Tree, 194
Indian Plum, 239
Indigofera, 221
Inula, 221
Inopsidium, 221
  acaule, 38
Ipheion, 221
Ipomoea, 287
Iresine, 287
Iris, 44, 126, 222
Iris, Alpine, 222
  Bearded, 126, 222
  Dutch, 222
  English, 222
  German, 126, 222
  Spanish, 222
  Water, 132, 222
Iron Tree, 241
Isatis, 222
Itea, 222
Ivy, 15, 94, 164, 217, 286
Ixia, 222

Ixiolirion, 222

Jacaranda, 287
Jacobinia, 287
Jacob's Ladder, 246
Japanese Bellflower, 245
Japanese Bitter Orange, 246
Jasmine, 154, 222, 287
  Box, 243
  Chilean, 289
Jasminum, 154, 222, 287
Jeffersonia, 222
Jerusalem Cherry, 300
Job's Tears, 199
Jovellana, 287
Jubaea, 287
Judas Tree, 196
Juglans, 222
Juncus, 132, 223
Juniper, 114, 223
Juniperus, 33, 144, 223
  chinensis, 114
  virginiana, 114

Kalanchoe, 273, 287
Kalmia, 166, 170, 223
Kangaroo Paw, 269
Kentia, 286
Kentucky Coffee Tree, 216
Kerria, 223
Kerria, White, 251
Kirengeshoma, 223
Kniphofia, 223
Knotweed, 246
Kochia, 287
Kolreuteria, 223
Kolkwitzia, 223

+ *Laburnocytisus*, 178, 223
Laburnum, 27, 105, 224
  anagyroides, 105, 106, 178
Lachenalia, 287
Lactuca, 224
Lady's Mantle, 182
Lady's Slipper, 280, 292
Laelia, 287
× *Laeliocattleya*, 287
Lagerstroemia, 287
Lagurus, 224
Lamarckia, 224

Index

Lamiastrum, 213
Lamium, 224
Lampranthus, 288
Lantana, 288
Lapageria, 288
Lapeirousia, 288
Larch, 27, 114, 224
  Golden, 248
Larix, 27, 114, 224
  decidua, 114
Larkspur, 204
Lathyrus, 49, 224
Laurel, Alexandrian,
  203
  Californian, 263
  Cherry, 70, 74, 248
  Portugal, 248
  Spotted, 188
Laurus, 224
Lavandula, 225
Lavateria, 225
Lavender, 225
Lavender Cotton, 253
Layia, 225
Lead Plant, 183
Leadwort, 195
Ledum, 225
Leek, 13, 14
Leiophyllum, 225
Lemna, 225
Lemon, 276
Lenten Rose, 218
Leonotis, 288
Leontopodium, 225
Leopard's Bane, 206
Leptospermum, 225, 288
Lespedesa, 225
Lettuce, 13
Leucocoryne, 288
Leucojum, 226
  autumnale, 226
Leucothoë, 226
Levisticum, 226
Lewisia, 39, 89, 226
Leycesteria, 226
Liatris, 226
Libertia, 226
Ligularia, 226
Ligustrum, 80, 227
  ovalifolium, 227
  vulgare, 117, 227
Lilac, 28, 117, 259

Lilium, 22, 43, 139, 227
  bulbiferum, 138
  'Enchantment', 138
  maculatum, 138
  sargentiae, 138
  speciosum, 138
  sulphureum, 138
  tigrinum, 138
Lily, 22, 138, 139, 227
Lily, African, 182
  Arum, 266, 303
  Blood, 285
  Boat, 297
  Climbing, 285
  Corn, 222
  Day, 218
  Giant, 193
  Ginger, 286
  Guernsey, 236, 291, 302
  Indian Shot, 274
  Peruvian, 183
  Plantain, 219
  St Bernard's, 185
  Scarborough, 302
  Toad, 262
  Voodoo, 299
  Wood, 262
  Zephyr, 267, 304
Lily of the Valley, 199
Lime (*Citrus*), 276
Lime, tree, 261
Limnanthus, 227
Limonium, 227
  latifolium, 227
Linanthus, 227
Linaria, 227
Linden, 261
Ling, 26, 191
Linum, 228
Lion's Ear, 288
Lippia, 228
Liquidambar, 228
Liquorice, 215
Liriodendron, 228
Liriope, 228
Lithops, 288
Lithospermum, 228
Living Stones, 288
Lobelia, 228, 288
  erinus, 288
Lobivia, 288
Lobularia, 228

Locust, 215, 251
Loganberry, 95, 156, 157,
  252
Lomatia, 229
Lonas, 229
Lonicera, 154, 229
  nitida, 80
Loofah, 289
Loosestrife, 230
Lophophora, 288
Loquat, 209
Loropetalum, 229
Lotus, 229
Lotus, 291
Lovage, 226
Love-in-a-mist, 236
Luculia, 289
Luffa, 289
Lunaria, 51, 229
  annua, 229
  rediviva, 229
Lungwort, 248
Lungwort, Smooth, 233
Lupin, 229
Lupin, Tree, 229
Lupinus, 20, 44, 45, 128,
  229
  arboreus, 229
Luzula, 230
Lycaste, 289
Lychnis, 230
Lycium, 230
Lycopodium, 289
Lyonia, 230
Lysichitum, 132, 230
Lysimachia, 230
Lythrum, 230

Maackia, 230
Macleaya, 230
Madwort, 183
Magnolia, 19, 33, 64, 68,
  166, 170, 231
  grandiflora, 231
Mahonia, 33, 34, 95, 231
  aquifolium, 231
Maidenhair Tree, 214
Maize, Ornamental, 303
Malcomia, 231
Mallow, 225, 231
  Greek, 256
  Jew's, 223

Poppy, 191
Shrubby, 218, 286
Malope, 231
Malus, 28, 31, 33, 103, 105, 231
sylvestris, 105, 106
Malva, 231
Mammillaria, 289
Mandevilla, 289
Manihot esculenta, 289
Manioc, 289
Maple, 15, 27, 28, 30, 180
Japanese, 64, 117, 180
Norway, 105
Maranta, 289
Marguerite, 275
Margyricarpus, 232
Marigold, 191
African, 301
French, 301
Marsh, 192
Marjoram, 13, 14, 238
Marvel of Peru, 290
Masdevallia, 289
Masterwort, 187
Matteuccia, 232
Matthiola, 232
bicornis, 232
Maxillaria, 289
Mazus, 232
Meconopsis, 20, 39, 40, 232
integrifolia, 38
Medinilla, 289
Medlar, 233
Melaleuca, 289
Melandrium, 232
Melia, 289
Melianthus, 290
Melissa, 232
Melittis, 232
Mentha, 233
Mentzelia, 233
Menyanthes, 233
Menziesia, 233
Mertensia, 233
Mesembryanthemum, 91, 290
Mespilus, 233
Metasequoia, 233
glyptostroboides, 81
Metrosideros, 290
Mexican Foxglove, 301

Mexican Orange
Blossom, 197
Michelia, 233, 290
Micromeria, 233
Mignonette, 250
Milium, 233
Milkweed, 187, 270
Milk Wort, 246
Millet, 233
Miltonia, 290
Mimosa pudica, 290
Mimulus, 234
Mind-your-own-business, 300
Mint, 233
Mint Bush, 296
Minuartia, 234
Mirabilis jalapa, 290
Miscanthus, 234
sinensis 'Zebrinus', 130
Mitchella, 234
Mitella, 234
Mitraria, 290
Mitrewort, 234
Mock Orange, 243
Molinia, 234
Moltkia, 234
Moluccella, 234
Monarch of the Veldt, 303
Monarda, 234
Monkey Flower, 234
Monkey Puzzle, 185
Monk's Hood, 181
Monstera, 290
deliciosa, 164
Montbretia, 202
Montia, 234
Moraea, 291
Morina, 234
Morisia, 234
Morning Glory, 287
Morus, 235
Mother of Thousands, 299
Mountain
Everlasting, 184
Mountain Heath, 244
Muehlenbeckia, 235
Mulberry, 235
Mullein, 264
Rosette, 249
Musa, 291

Muscari, 42, 235
Mutisia, 235
Myosotidium, 235, 291
Myosotis, 51, 235
Myrica, 235
Myriophyllum, 235
Myrrh, 235
Myrrhis, 235
Myrtle, 14, 236
Myrtle, Bog, 235
Grape, 287
Sand, 225
Wax, 235
Myrtus, 236

Nandina, 236
Narcissus, 42, 136, 236
Nasturtium, 47, 263
Navelwort, 238
Nectarine, 105, 248
Neillia, 236
Nelumbo, 291
Nemesia, 291
Nemophila, 236
Neoregelia, 291
Nepenthes, 291
Nepeta, 236
Nephrolepis, 291
Nerine, 236, 291
Nerium, 291
oleander, 101
Nertera, 236
New Zealand Flax, 243
Nicandra, 236
Nicotiana, 291
Nidularium, 291
Nierembergia, 236
repens, 236
Nigella, 236
Nomocharis, 237
Nothofagus, 237
Notholirion, 237
Notocactus, 291
Notospartium, 237
Nuphar, 237
Nymphaea, 133, 237, 292
Nymphoides. 237
Nyssa, 237

Oak, 26, 33, 249
Oat, 188
Ochna, 292

Index

Odontoglossum, 292
Oenothera, 237
Olea, 292
Oleander, 291
Olearia, 238
Olive, 13, 14, 292
Omphalodes, 238
Oncidium, 292
Onion, 14
  Ornamental, 182
  Tree, 138
Onoclea, 238
Ononis, 238
Onopordon, 238
Onosma, 238
Ophiopogon, 238
Ophrys, 238
Oplismenus, 292
Opuntia, 292
Orach 13
Orange, 276
Orchid, Bee, 238
  Lady's Slipper, 203
  Marsh, 238
  Moth, 294
  Pyramidal, 238
  Spotted, 238
Orchis, 238
Oregon Grape, 231
Origanum, 238
Ornithogalum, 239, 292
Orontium, 239
Osmanthus, 239
× *Osmarea*, 239
Osmaronia, 239
Osmunda, 239
Oso Berry, 239
Osteomeles, 239
Ostrowskia, 239
Ostrya, 239
Oswego Tea, 234
Othonnopsis, 239
Ourisia, 239
Oxalis, 239, 292
Oxydendrum, 240
Ozothamnus, 240

Pachysandra, 240
Paeonia, 126, 240
Painted Tongue, 298
Paliurus, 240
Palm, 275, 286, 287, 295

Chusan, 262
Coconut, 276
Date, 13, 295
Pancratium, 240
Pandanus, 292
Pandorea, 292
Panicum, 240
Pansy, 102, 265
  Summer Flowering, 265
  Winter/Spring
    Flowering, 265
Papaver, 240
  alpinum, 38
  orientale, 84, 128, 240
Paphiopedilum, 292
Paraquilegia, 241
Parochetus, 241
Parodia, 293
Parrotia, 241
Parthenocissus, 81, 241
Partridge Berry, 234
Pasque Flower, 249
Passiflora, 94, 154, 241, 293
Passion Flower, 94, 154,
  241, 293
Paulownia, 241
Pavonia, 293
Peach, 103, 105, 248
Pear, 14, 103, 105, 106,
  249
Pearl Fruit, 232
Pearlwort, Heath, 253
Pearly Everlasting, 184
Pedilanthus, 293
Pelargonium, 176, 293
  × *domesticum*, 293
  × *hortorum*, 293
  peltatum, 293
  Ivy-leaf, 293
  Regal, 293
  Zonal, 293
Pellaea, 293
Pellionia, 293
Peltiphyllum, 241
Pennisetum, 241
Penny Cress, 261
Penstemon, 241
Pentas, 293
Peony, 240
  Tree, 16, 240
Peperomia, 89, 294
Pepper, Ornamental, 274

Pepper Elder, 294
Pereskia, 294
Perilla, 294
Periploca, 242
Periwinkle, 265
Pernettya, 34, 36, 157, 166,
  242
Perovskia, 242
Persea, 294
Petasites, 242
Petrophytum, 242
Petunia, 294
Phacelia, 242
Phaius, 294
Phalaenopsis, 294
Phalaris, 242
Phellodendron, 242
Philadelphus, 81, 172, 243
Phillyrea, 243
Philodendron, 152, 164, 294
Phlomis, 243
Phlox, 243, 294
  drummondii, 294
  paniculata, 84, 243
Phoenix, 295
Phormium, 243
Photinia, 243
Phuopsis, 243
Phygelius, 243
Phyllitis, 244
Phyllodoce, 244
Phyllostachys, 244
Physalis, 244, 295
Physocarpus, 244
Physostegia, 244
Phyteuma, 244
Phytolacca, 244
Picea, 244
  abies, 114
  pungens, 114
  pungens glauca, 114
Pick-a-back Plant, 163,
  261, 302
Pickerel Weed, 247
Pieris, 166, 170, 244
Pilea, 295
Pileostegia, 245
Pimpinella, 245
Pine, 27, 114, 245
  Norfolk Island, 270
  Scots, 26
  Umbrella, 255

Pineapple, 269
  Flower, 210
Pinguicula, 245, 295
Pink, 20, 100, 101, 128,
  160, 204
Pinus, 27, 114, 245
  sylvestris, 26, 114
Piptanthus, 34, 245
Pistia, 295
Pitcairnia, 295
Pitcher Plant, 291, 298
  Californian, 204
Pittosporum, 245, 295
Plane, 245
Plantago, 245
Plantain, 245
Plantanus, 245
Platycerium, 295
Platycodon, 245
Platystemon, 245
Plectranthus, 295
Pleione, 295
Plum, 103, 105, 248
  Rootstock 'Brompton',
    105, 106
  Rootstock 'Myrobalan
    B', 105, 106
Plumbago, 295
Plumeria, 296
Poa, 245
Poached Egg Flower, 227
Podocarpus, 246
Podophyllum, 246
Poinsettia, 283
Pokeweed, 244
Polemonium, 246
Polianthes, 296
Polka Dot Plant, 286
Polyanthus, 51, 52, 247
Polygala, 246
Polygonatum, 246
Polygonum, 246, 250
  baldschuanicum, 246
Polypodium, 246, 296
Polypody, 246
Polyscias, 296
Polystichum, 246
Pomegranate, 14, 296
Poncirus, 246
Pondweed, 247
Pontederia, 247
Poor Man's Orchid, 299

Poplar, 15, 80, 82, 247
Poppy, 38, 240
  Blue, 232
  Californian, 210
  Californian Tree, 251
  Horned, 215
  Iceland, 240
  Oriental, 240
  Plume, 230
  Prickly, 186
  Shirley, 240
  Tree, 204
  Welsh, 232
Populus, 15, 80, 247
Portulaca, 296
Potamogeton, 247
Potentilla, 34, 172, 147
Pouch Flower, 286
Pratia, 247
Prickly Heath, 242
Prickly Pear, 292
Primrose, 126, 247
  Drumstick, 247
Primula, 20, 39, 45, 51,
  126, 132, 247, 296
  denticulata, 84, 126,
    247
  edgworthii, 39
  gracillipes, 39
  × *kewensis*, 296
  malacoides, 296
  obconica, 296
  reidii, 38
  sinensis, 296
  sonchifolia, 39
  × *variabilis*, 247
  viali, 38
  whitei, 39
Privet, 80, 117, 227
Prophet Flower, 186
Prostanthera, 296
Protea, 296
Prunella, 247
Prunus, 31, 33, 70, 103,
  105, 248
  avium, 105, 106
  glandulosa, 248
  pumila, 248
  laurocerasus, 248
  lusitanica, 248
Pseudolarix, 248
Pseudopanax, 296

Pseudotsuga, 248
Ptelea, 248
Pteris, 60, 296
Pterocarya, 248
Pulmonaria, 126, 248
Pulsatilla, 249
  vulgaris, 249
Punica, 296
Purslane, 13
Puschkinia, 249
Pyracantha, 33, 172, 249
Pyrethrum, 126, 249
Pyrus, 103, 105, 249

Quamash, 192
Queen of the Night, 299
Quomoclit, 297
Quercus, 249
  robur, 26, 27
Quince, 202
  Malling A, 105, 106
  Ornamental, 196

Radish, 13
Ramonda, 89, 249
Ranunculus, 250
  aquatilis, 250
  lingua, 250
Raoulia, 131, 250
Raspberry, 136, 252
Rebutia, 297
Red-hot Poker, 223
Redwood, 255
  Dawn, 233
Reed, 132
Reedmace, 263
Rehmannia, 297
Reinwardtia, 297
Reseda, 250
Reynoutria, 250
Rhamnus, 250
  cathartica, 26
Rhaphiolepis, 250, 297
Rhazya, 250
Rheum, 250
Rhipsalidopsis, 297
Rhipsalis, 297
Rhododendron, 19, 27, 28,
  34, 36, 40, 64, 68, 109,
  110, 112, 166, 170, 250,
  297
  Hardy Hybrids, 109, 250

Index

Rhododendron (cont.)
Javanese, 297
ponticum, 100, 112
simsii, 297
Rhodohypoxis, 251
Rhodotypos, 251
Rhoeo, 297
Rhoicissus, 297
Rhubarb, 250
Rhus, 85, 251
typhina, 84, 136
Ribbon Wood, 219
Ribes, 78, 81, 251
grossularia, 251
nigrum, 251
sativum, 251
Ricinus, 297
Robinia, 27, 34, 251
pseudoacacia, 105, 106
Rochea, 298
Rock Cress, 185
Jessamine, 184
Purslane, 273
Rose, 217
Rodgersia, 251
Romneya, 251
coulteri, 86, 126
Rosa, 33, 119, 252
canina, 120
canina 'Inermis', 120
canina 'Pfander', 120
dumetorum 'Laxa', 120
rugosa 'Hollandica', 121
Roscoea, 252
Rose, 14, 16, 19, 28, 119, 175, 252
Dog, 120
Climber, 119, 252
Floribunda, 119, 252
Hybrid Tea, 119, 252
Miniature, 252
Rambler, 119, 252
Shrub, 119, 252
Rosmarinus, 252
Rosemary, 252
Bog, 184
Rubber Plant, 95, 152, 283
Rubus, 95, 136, 144, 156, 252
cockburnianus, 252
odoratus, 252
parviflorus, 252

spectabilis, 252
Rudbeckia, 44, 252
Rue, 14, 252
Goat's, 213
Meadow, 260
Ruellia, 298
Ruscus, 253
Rush, 132, 223
Flowering, 191
Ruta, 253

Sage, 253, 298
Scarlet, 298
Sagina, 253
subulata 'Aurea', 253
Sagittaria, 253
St John's Wort, 220
Saintpaulia, 89, 135, 298
Salix, 80, 82, 101, 253
Sallow, 253
Salpiglossis, 298
Salvia, 253, 298
horminum, 253
splendens, 298
Sambucus, 253
Sandwort, 186
Sanguinaria, 253
Sanguisorba, 253
Santolina, 253
Sansevieria, 91, 135, 298
trifasciata, 91
trifasciata 'Laurentii', 298
Sanvittalia, 254
Saponaria, 254
Sarcococca, 254
Sarracenia, 298
Sassafras, 254
Satin Flower, 256
Sauromatum, 299
Savory, 13
Saxifraga, 131, 254, 299
sarmentosa, 163
stolonifera, 163, 299
Saxifrage, 131, 254
Scabiosa, 254
Scabious, 254
Giant, 195
Schefflera, 299
Schizandra, 254
Schizanthus, 299
Schizophragma, 254
Schizostylis, 255

Schlumbergera, 299
Sciadopitys, 255
Scilla, 42, 255
Scindapsus, 299
Scirpus, 255
Screw Pine, 292
Scrophularia, 255
Scutellaria, 255
Sea Buckthorn, 219
Sea Holly, 209
Sea Lavender, 227
Sedge, 193
Sedum, 91, 255, 299
Selaginella, 299
Selenicereus, 299
Self Heal, 247
Sempervivum, 131, 255
Senecio, 255, 300
cineraria, 300
× *hybridus*, 300
Sensitive Plant, 290
Sequoia, 255
Sequoiadendron, 256
Setcreasea, 300
Shaddock, 276
Shamrock Pea, 241
Shoo-fly, 236
Shooting Stars, 206
Shortia, 256
Shrimp Plant, 271
Sidalcea, 256
Silene, 256
Silk Vine, 242
Silver Fir, 180
Silybum, 256
Sinningia, 89, 300
Sisyrinchium, 256
Skimmia, 33, 256
Skull Cap, 255
Skunk Cabbage, 230
Slipperwort, 191, 273
Smilacina, 256
Smithiantha, 300
Snapdragon, 270
Sneezeweed, 217
Snowberry, 259
Snowdrop, 137, 213
Tree, 216
Snowflake, 226
Snow in Summer, 195
Snow on the
Mountain, 211

Soapwort, 254
Solanum, 256, 300
  capsicastrum, 300
  pseudocapsicum, 300
Soldanella, 256
Soleirolia, 300
Solidago, 256
× *Solidaster*, 257
Solomon's Seal, 246
Sophora, 257
Sophronitis, 300
Sorbaria, 257
Sorbus, 31, 33, 257
  aria, 26, 105, 106
  aucuparia, 105, 106
Sorrel, Wood, 239
  Tree, 240
Sparaxis, 257
Sparmannia, 300
Spartium, 34, 37, 257
Spathiphyllum, 300
Speedwell, 264
Spider Flower, 276
Spider Plant, 162, 275
Spiderwort, 262
Spindle Tree, 210
Spiraea, 81, 172, 257
Spleenwort, 187, 271
Sprekelia, 300
Spruce, 114, 244
  Blue, 114
  Norway, 114
Spurge, 210
  Mountain, 240
  Slipper, 293
Squill, 249, 255
Stachys, 257
Stachyurus, 257
Stapelia, 301
Staphylea, 257
Star Cluster, 293
Star of the Veld, 281
Stephanandra, 258
Stephanotis, 301
Sternbergia, 258
Stewartia, 258
Stipa, 258
Stock, 232
  Beauty of Nice, 232
  Brompton, 232
  East Lothian, 232
  Night Scented, 232

Ten-week, 232
  Virginian, 231
Stokes' Aster, 258
Stokesia, 258
Stonecrop, 255, 299
Stork's Bill, 209
Stranvaesia, 258
Stratiotes, 258
Strawberry, 20, 150, 158, 212
  Tree, 185
Straw Flower, 217
Strelitzia, 301
Streptocarpus, 89, 301
Streptosolen, 301
Styrax, 258
Sumach, 251
Summer Cypress, 287
Sundew, 207, 281
Sunflower, 217
Sweet Flag, 181
Sweet Gum, 228
Sweet Pea, 49, 224
Sweet Rocket, 218
Sweet William, 51, 205
Swiss Cheese Plant, 290
Sycamore, 27, 105
Sycopsis, 279
Symphoricarpos, 33, 136, 259
Symphyandra, 259
Symphytum, 259
Symplocos, 259
Syngonium, 301
Syringa, 28, 117, 259

Tagetes, 301
Tamarisk, 259
Tamarix, 259
Tanacetum, 260
Tapioca, 289
Taxodium, 260
Taxus, 33, 114, 260
  baccata, 26, 114
Teasel, 205
Tecophilaea, 260
Tellima, 260
Telopea, 301
Tetranema, 301
Teucrium, 260
Thalictrum, 260
  'Hewitt's Double', 260

Thelypteris, 260
Thermopsis, 261
Thistle, 14
  Globe, 207
  Milk, 256
  Scotch, 238
Thlaspi, 261
Thorn, 105, 201
Thrift, 186
Throatwort, 262
Thunbergia, 301
Thuja, 261
Thujopsis, 261
Thyme, 131, 261
Thymus, 131, 261
Tiarella, 261
Tibouchina, 302
Tick Seed, 199
Tidy Tips, 225
Tiger Flower, 261
Tiger's Jaw, 283
Tigridia, 261
Tilia, 261
Tillandsia, 302
Toadflax, 227
Tobacco Plant,
  Flowering, 291
Tolmiea, 21, 261, 302
  menziesii, 163, 302
Tomato Tree, 280
Torenia, 302
Torreya, 261
Trachelium, 262
Trachelospermum, 262
Trachycarpus, 262
Trachystemon, 262
Tradescantia, 101, 164, 262, 302
Tragopogon, 262
Traveller's Joy, 28
Tree of Heaven, 182
Trefoil, 262
Trichocereus, 302
Tricyrtis, 262
Trifolium, 262
Trillium, 262
Tritonia, 262
Trochodendron, 262
Trollius, 263
Tropaeolum, 47, 263, 302
  peregrinum, 302
Tsuga, 263

Index

Tuberose, 296
Tulip, 136, 137, 263
Tulip Tree, 228
Tulipa, 42, 136, 263
Tupelo Tree, 237
Turnip, 13
Turtlehead, 196
Typha, 132, 263

Ulex, 263
  europaeus, 26
Ulmus, 263
  procera, 15
Umbellularia, 263
Umbrella Plant, 241, 279
Ursinia, 302
Utricularia, 263
Uvularia, 263

Vaccinium, 157, 264
Valerian, 195, 264
Valeriana, 264
Vallota, 302
Vancouveria, 264
Vanda, 135, 303
Veltheimia, 303
× *Venidio-arctotis*, 303
Venidium, 303
Venus Fly Trap, 281
Veratrum, 264
Verbascum, 264
Verbena, 264, 303
Verbena, Lemon, 228
Vernonia, 264
Veronica, 264
Veronica, Shrubby, 217
Vervain, 264, 303
Viburnum, 33, 34, 81, 172, 265
  opulus, 26
  plicatum, 144
Vinca, 265
Vine, 97, 154, 265
  Chilean Glory, 281

Cup and Saucer, 276
Grape, 14, 17, 97, 265
  Russian, 246
  Wonga-wonga, 292
Viola, 20, 45, 102, 265
Violet, 265
Violet Cress, 221
  Dog's Tooth, 209
  Water, 219
Viper's Bugloss, 207
Virginia Creeper, 241
Viscaria, 265
Vitis, 154, 265
  'Brant', 97
  coignetiae, 97
  davidii, 97
  pulchra, 97
  vinifera, 97
  'Black Hamburg', 97
  'Buckland Sweet-
    water', 97
  'Purpurea', 97
Vriesia, 303

Wahlenbergia, 265
Waldsteinia, 265
Wallflower, 18, 51, 52, 196
Walnut, 222
Wand Flower, 205
Wandering Jew, 302, 304
Water Lettuce, 295
Water Lily, 133, 237
  Fringed, 237
  Tropical, 292
  Yellow, 237
Water Milfoil, 235
Water Soldier, 258
Water Starwort, 191
Watsonia, 303
Wattle, 268
Wax Flower, 286, 301
Weigela, 81, 172, 266
Wellingtonia, 256

Whitebeam, 26, 31, 105, 257
Willow, 80, 82, 101, 253
Willow Herb, 208
Windflower, 184
Wingnut, 248
Winter Aconite, 208
Winter Cherry, 300
Winter Heliotrope, 242
Winter Sweet, 197
Wire Plant, 235
Wisteria, 81, 154, 266
Witch Hazel, 19, 216
Woad, 222
Woodruff, 187
Woodrush, 230
Woodwardia, 266, 303
  radicans, 303
  virginiana, 266
Wormwood, 186
Woundwort, 257

Xeranthemum, 266

Yarrow, 181
Yew, 26, 114, 260
  Plum, 195
Yellow Archangel, 213
Yucca, 136, 266

Zantedeschia, 266, 303
  aethiopica, 266
Zanthoxylum, 267
Zauschneria, 267
Zea mays, 303
Zebrina, 304
  pendula, 101, 164
Zelkova, 267
Zenobia, 267
Zephyranthes, 267, 304
Zinnia, 304
Zinnia, Creeping, 254